THE NEW MIDDLE CLASS AND DEMOCRACY IN GLOBAL PERSPECTIVE

The New Middle Class and Democracy in Global Perspective

Ronald M. Glassman
Professor of Sociology
William Paterson College of New Jersey
Wayne
New Jersey

First published in Great Britain 1997 by
MACMILLAN PRESS LTD
Houndmills, Basingstoke, Hampshire RG21 6XS and London
Companies and representatives throughout the world

A catalogue record for this book is available from the British Library.

ISBN 0–333–68305–6

First published in the United States of America 1997 by
ST. MARTIN'S PRESS, INC.,
Scholarly and Reference Division,
175 Fifth Avenue, New York, N.Y. 10010

ISBN 0–312–17421–7

Library of Congress Cataloging-in-Publication Data
Glassman, Ronald M.
The new middle class and democracy in global perspective / Ronald
M. Glassman.
p. cm.
Includes bibliographical references (p.   ) and index.
ISBN 0–312–17421–7 (cloth)
1. Democracy. 2. Middle class. 3. Capitalism. 4. High
technology—Social aspects. 5. Bureaucracy. I. Title.
JC423.G58 1997
321.8'09'049—dc21   96–46503
                                                          CIP

This book is printed on paper suitable for recycling and made from fully managed and
sustained forest sources.

10   9   8   7   6   5   4   3   2   1
06 05 04 03 02 01 00 99 98 97

Printed and bound in Great Britain by
Antony Rowe Ltd, Chippenham, Wiltshire

Contents

Introduction: Industrial Capitalism and Legal-Representative Democracy: The End of History?

THE MAIN LINE OF HISTORICAL DEVELOPMENT

We have been witnessing, from the 1970s onward, the collapse of both fascism and communism. In Spain, fascism slowly eroded and then suddenly gave way to legal-parliamentary government and common-market capitalism. And, of course, the world was as stunned by the fall of Soviet communism as it had been by the Russian revolution.

Fascism, in Italy, Germany and Japan, was defeated and dismantled after World War II, while in Spain, the transition from fascism to capitalist democracy occurred surprisingly smoothly, as the success of the European common market overwhelmed all opposing ideologies and swept Spain into its upward spiral.

It should be noted, however, that Germany and Japan retained their tradition of government involvement in the industrial productive process. This phenomenon must be carefully analysed, in post-fascist terms, for its implications in both the economic structure and the political system of Germany and Japan. Since Germany has become the model for all Europe, including Eastern Europe, and, since Japan is rapidly becoming the model for all of Asia (including China, perhaps), the implications of their post-fascist economic and political systems could be profound.

By this latter statement, we do not mean to imply that the fascist movement is still alive. Murderous, terroristic totalitarian fascism has been eliminated in Germany and Italy, and, of course, it never existed in Japan (wherein the fascism was more of a traditional militaristic authoritarian phenomenon). Fascism, as a mass movement with charismatic leadership, and as an ideological alternative to capitalist democracy, and as a totalitarian system of government, is dead – though neo-fascist parties still haunt the European democracies with their nationalist and racist doctrines.

In the Third World, of course, neo-fascist dictatorships exist, and will continue to exist. These dictatorships are not considered a model for modern

1

government but rather as temporary expedients existing to create 'order' within societies racked by poverty and anarchy during their attempted industrialization period.

Communism has also become a non-model for modernization. Once, it had been the great hope of the Third World, promising the possibility of 'leaping over stages' and achieving rapid industrialization. Communism has not been able to achieve this goal. It has not been a complete failure – Russia and China, after all, are world powers with modern military capacities and a modern educational and economic infrastructure. Vietnam and Cuba also developed powerful military capacities and the beginnings of a modern infrastructure.

However, the communist party state, with its byzantine administrative bureaucracy and total disregard for market dynamics, and enlightenment political principles, has proven to be a colossal impediment to industrial modernization and political democratization. Once created and legitimated, it is so difficult to remove that neither the charismatic Mao, during his Cultural Revolution, nor the liberal-democratic Gorbachev, during his *glasnost–perestroika* period – nor the embattled Fidel Castro faced with the near-collapse of his Cuban dream – were able to dismantle it. It took massive protests, and/or bloodshed and revolution in Eastern Europe to eliminate the communist leviathan, and even now, the Eastern European economies are suffering in their attempted transition away from communism. A different, but also traumatic transition, is occurring in China, and will undoubtedly occur in Cuba as well.

The Third World nations have been witnessing all of this, and have become soured on communism as an alternative system to capitalism. Communism, as the great hope for the underdeveloped world, is now viewed as a fantasy.

Finally, with the recent success of the Four Tigers – Hong Kong, Taiwan, Singapore and South Korea – capitalism and legal-representative democracy have gained a new legitimation for successful modernization in developing nations. Be forewarned that it is the capitalism that has been dramatically successful in the Pacific Rim, with the democracy limping along and often overridden by authoritarian or military governments. Nonetheless, the success of the new Pacific Rim capitalism has been truly phenomenal (with Thailand and Indonesia now joining the Four Tigers), and the democracy, in some of these nations, has been steadily improving – especially in Taiwan, South Korea, and Hong Kong, with Singapore still authoritarian.[1]

Thus, with communism more discredited every day – the Russian economic collapse putting the nail in the coffin – and with fascism dead as a mass movement, tolerated only where anarchy and anomie engender its emergence, capitalism and legal-representative democracy have truly become the main line of historical development.

Francis Fukayama[2] has gone so far as to call capitalist democracy the 'end of history', parodying Marx on the stage of pure communism being the end of history. Of course, history never ends, and it has no teleological goal to reach. However, Fukayama is correct that for the near future, capitalist-democratic society has become the model for emulation – the standard to which all nations aspire.

If capitalist democracy is the main line of history, the questions that remain are: just what is the structure of today's high technology capitalism, and, how will legal-representative democracy function as undergirded by this new form of capitalism?

DIFFERENT FORMS OF CAPITALISM, DIFFERENT FORMS OF DEMOCRACY

In past historical eras, differing forms of capitalist economic systems were linked with very different forms of democracy. Commercial capitalism or trade capitalism undergirded the direct democracy of the Greek city-states, and the revival of city-state commerce in Renaissance Italy and post-feudal Germany brought direct democracy back to the centre of history once again.[3]

As commercial capitalism spread to the countryside in England, France and the Germanic countries, the democratic assemblies of the city-states were extended through the institution of the *standestaat*[4] – or estate-state – into a representative system that extended its authority over a larger territory than was possible with the direct democratic assemblies of the city-states.

However, though commercial capitalism and its new representative-democratic institution – the parliament – did begin to triumph over the kingly-feudal-peasant agrarian institutions of Europe, neither capitalism nor democracy became firmly stabilized or fully legitimated until the industrial revolution.[5]

Only in the United States and Holland did representative democracy fully function on a commercial-capitalist base.[6] England had a property qualification and was essentially oligarchic; France was so divided it remained anarchic; while the rest of Europe retained monarchical-feudal forms of government.

But, the industrial revolution was created by commercial capitalists applying machine technology to the (previously artisan) production process. Whether this application of the machine to the productive process was protestant-engendered or not,[7] it was capitalist in economic orientation. Thus, the industrial revolution was a capitalist-industrial revolution, which not only revolutionized the production of goods, but altered the entire social structure toward capitalist economic, legal, cultural and political institutional configurations.[8]

Firmly undergirded by these capitalist-industrial institutions and ideas, legal-representative democracy emerged as a viable form of government for the nation-states that replaced the kingly states and the city-states as the new political entity of the modern world.

As we move now from industrial capitalism to high-technology industrial capitalism, we wonder, once again, whether the form of democracy will change as the form of capitalism changes.

Will high-technology capitalism engender an alteration in legal-representative democracy as dramatic as the change from direct democracy to parliamentary democracy? The new communications media alone, which have already been engendered by the high technology revolution, could produce a change in the process of modern democracy, and may even engender a change in its form. Already one hears debates about instant referenda, conducted through computer–television hookups,[9] inter-active town meetings conducted through computer–telephone combinations,[10] and much more.

Theorists of the trans-national[11] or multi-national corporations are even debating whether the nation-state will survive as high-technology capitalism becomes more and more global in its organizational structure.

Let us reiterate once again, then, that though democracy survived the transition from commercial capitalism to industrial capitalism, its form and process certainly changed. City-state direct democracy was a very different institution than the legal-representative democracy of the mass nation-state. Yet Rousseau[12] was wrong: legal-representative democracy is democratic, and, if less democratic than the assemblies of the city-states, it has been more stable.

Will democracy survive the new transition from industrial to high-tech industrial capitalism? Will its form and process change again, if it does survive? Will democracy need institutional bolstering in order to remain effective, as Dahl[13] and others[14] are insisting? Will democracy be overwhelmed and pushed aside by the technocratic-bureaucratic management teams that drive the modern productive process, as Weber[15] and Mills[16] have warned?

In another vein, will variations in the form of legal-representative democracy become significant? That is, will a presidential system, such as that of the USA, be more or less problematic than a prime-ministerial system? Is it good that Boris Yeltsin is elected separately from the Russian legislature, or is it dangerous? Is presidential 'charisma' a dynamic and energizing factor that in the age of technocratic bureaucracy[17] could be useful for the proper functioning of electoral democracy, or is the dictatorial potential too dangerous?

All these questions must be addressed before we can languish in any 'end of history' euphoria.

Furthermore, in terms of the economic system we must ask: just what is the structure of high-technology industrial capitalism? This new mode of production has so recently emerged that its exact nature has remained controversial.

Marx, already, in *Das Capital*,[18] was describing an industrial system quite different from that seen by Smith a century earlier. Smith's pin-factory[19] was a far cry from the huge heavy-industrial factories Marx witnessed, and, of course, Marx never lived to see the steel and auto factories of the early twentieth century, or the automated 'producer units' of today.

Other economic theorists began to observe changes in the institutional structure of the productive system that had nothing to do with the capitalist market dynamics that continued to motivate the industrial enterprises. Veblen, for instance, keenly noted the beginnings of a separation in the structure of industrial production from that of finance capitalism. He did not suggest that business, or market principles, would not continue to motivate the system, but rather that the hands-on processes of production were being improved and operated by scientifically and technologically sophisticated 'engineers', rather than by the businessmen-financiers.[20]

Veblen focused the attention of the economists on the technology of production, and did down-play the importance of the finance capitalists. However, the 'financiers' have turned out to be equally as important as the 'engineers' in the production process – just look at the communist systems if you doubt the importance of market dynamics in efficient production, and in technological innovation itself.

Yet, market economists themselves have been divided in their description and analysis of the modern economic system. The neo-classicists, such as Friedman,[21] present one picture, while Keynes[22] and his followers present another. In terms of the role of government in investment, interest rates, stock market and banking regulations and every other aspect of market functioning, the two schools of thought differ. To make the situation even more complex, in Europe and Japan, a more government-oriented approach than that of the Keynesians predominates, presenting us with a third theoretical orientation derived from List[23] and Schumpeter,[24] grounded in the historical experience of trying to catch up with England and the USA.

And still the complications continue: Galbraith,[25] Bell,[26] Reich,[27] and earlier, Myrdahl,[28] have been attempting to describe the exact institutional structure of this elusive, evolving modern economic system. Nobody has quite got it right yet. Galbraith focuses on the role of the state, Bell on the new service sector, Reich in the transnational nature of production.

The competing theorists have not even agreed on a name for the new economic system – that is why we are using the clumsy designation: high-technology industrial capitalism.

The new mode of production utilizes computers, robots, global information networks, and other scientifically advanced technologies for the mass production of technologically sophisticated products and services. This we know, but we are not sure what form the organizational structure surrounding such production will take. The structure is unclear because it is still evolving.

It is also unclear because there are three variations emerging: the Japanese economic system is different from the American, and different again from the German and Scandinavian models. Since these systems are different from one another, and since it is difficult to study an incomplete, rapidly evolving process, it is even more problematic to establish the impact on society (and on the political process) that the new economic system will eventually have.

THREE DIFFERENT MODELS OF HIGH-TECHNOLOGY INDUSTRIAL CAPITALISM

We shall analyse the economic system in detail in the next chapter, and then describe the class structure that has begun to emerge from it. Here, we wish to underscore that there are three different forms of high-technology industrial capitalism which have evolved: firstly, the more purely capitalist system of the United States, England, and Canada; second, the post-fascist or 'corporatist' model of Germany and Japan, characterized by heavy government involvement; and third, the democratic-socialist model of the Scandinavian countries – with Germany and the other European Union countries evolving toward the Scandinavian model in terms of social programmes – but with the Scandinavian countries evolving toward the German model in terms of industrial planning and corporate–government partnership.

All three models, it must be emphasized, are market-oriented, the communist-style non-market, dictatorial-bureaucratic, 'command-economy' model having died of inefficiency and lack of productivity.

It is crucial to highlight the market nature of all three systems, because of the colossal failure of the communist economic systems which tried to operate without market principles: production quotas were filled for specified products without any regard for efficiency, profit, prices, supply and demand, or any other market mechanisms. The very visible hand of the centralized party bureaucracy controlled every aspect of industrial production. Since this communist party 'command' economy failed, it will not likely serve as a model for future societies. Thus, capitalist market-oriented systems will undoubtedly prevail and lead us through the next epochs of history.

However, the variations between the three systems may prove important for future historical development – both in terms of economic structure and political structure. For instance, it is now becoming clear that the Japanese system is very different from the more classically oriented American system. The key difference lies with the Japanese government bureaucracy acting as an advisory agency.[29]

There are a number of Japanese government agencies – the most famous of which is MITI[30] – which intervene directly in the economy. They back certain companies and withdraw support from others; they control the flow of capital from the banks; they set prices differently within Japan than in foreign nations (the prices are higher for the Japanese); they restrict and control trade.

Government bureaus in cooperation with the Japanese giant business conglomerates – which used to be called Zaibatsu and are now called Kieretsu – together control the economic system. Banks and the stock market are tightly controlled by this alliance of bureaucrats and businessmen.

Thus, though the Japanese system utilizes market principles quite thoroughly – in terms of every principle enunciated by the classical economists, except for *laissez-faire*! – one would have to call this system a 'guided' market,[31] or even a 'governed' market.[32]

We shall describe this Japanese system in detail in the next chapter. Here, what is important to understand is that the Japanese government bureaucracy not only intervenes to guide the economy, but also to guide the polity.

That is, the government bureaucracy overlaps directly with the Japanese ruling party and with all of the Japanese parliamentary leaders. Parliamentary decisions – even the choice of the prime minister and cabinet members – are directly controlled by the government bureaucracy. Often, the government bureaucrats designate certain individuals to run for office from specified political districts.[33]

Political scientists are not sure whether the Japanese system is really a parliamentary democracy or not.[34] No opposition party has been able to hold power, and few opposition candidates have become important polit-ical figures. However, there are elections, there are law courts, and the people do occasionally assert themselves at the polls, and in 1994–95 a socialist prime minister did come to power.

We shall have to analyse this Japanese model very carefully, in terms of both its economic and political implications for two reasons: firstly, the Japanese model of high-technology capitalist production has become the model for most of Asia, including South Korea, Taiwan, Singapore, Thailand, Indonesia, and now, perhaps, China and Vietnam; second, the success of the Japanese economy may force the Americans and Europeans to emulate it. If this latter occurs, the modified market system could lead to modifications in the political system.

The post-fascist economic system of Germany is already similar to that of Japan in certain institutional ways. For instance, the government and the corporations work more closely together than is the case in the Anglo-Saxon countries,[35] and the banks are more tightly controlled, as in Japan.

It is possible that the Swedish economic planning board and its German counterpart will become more and more like MITI in their power relation-ships with the giant corporations and the parliament. What will this mean for democracy? Already in 1972 when we visited Sweden, the guide jok-ingly said, 'I'll take you to the planning board offices instead of the par-liament chamber, because they have the real power anyway.' In terms of political theory on the separation of power and the *laissez-faire* state, this is no joking matter.

History does not end, and its 'main line' may not be the simple distance between two points that Fukayama suggested.

A COMPARISON WITH THE 'AGE OF KINGS'

Step back for a moment, away from the modern era, to the 'age of kings'. Reinhard Bendix, for instance, in *Kings or People*,[36] describes the era of kings and then the evolution of democratic government in Greece, Rome and post-feudal Europe. The age of kings can be thought of as one of those great chunks of human history in which the world's political structure was characterized by the domination of the kingship.

But Marx already distinguished between 'oriental' kingship and Greek and Roman 'slave' societies.[37] In the 'age of Kings', the Greeks and Romans really didn't have any.[38] When the Romans finally attempted to establish an emperorship, it ended up as a non-legitimate institution, overthrown regularly by the army and the senate, and never really 'divine'.

The Spartans were supposed to have two kings, but they really had none – they actually had two generals, who ruled in concert with the council of elders.[39] The ancient Jews very reluctantly established a kingship, never fully legitimating it, and constantly railing against it from their pastoral hillsides.[40]

Many variations existed even among the societies wherein kingship did gain legitimation and longevity. The Egyptian Pharaohs ruled with great authority and stability, supported by the priesthood and truly made 'divine' in the minds of the people; whereas, because of the constant military invasions, the kingships of Mesopotamia never really gained stability, even though they were considered legitimate.[41] Kings, in the Fertile Crescent, were not only overthrown by military conquest, but by competing factions within the upper strata, all the time. Brothers against brothers, sons against fathers, one aristocratic clan against another – and then an invasion from one of the myriad military pastoralists – made kingly stability impossible in the land of the Tigris and Euphrates rivers.

China, like Egypt, produced divine stability for its kings, yet even here there is a variation. For, though the Egyptian bureaucrats, the nomarchs, were excellent administrators, they never gained real power in Egypt. Yet, the Chinese mandarins, at times of weak emperors, often gained nearly full power within the Chinese kingly state.[42]

Thus, in the 'age of kings', there were some key societies with no kings at all, others with unstable kingships, and still others with non-legitimate kings. Yet, we tend to view this era of history as if all the kingships were 'divine' and stable, and all of history was the history of kings. So focused on kings did historians become that they tended to view the period of tribal gerontocracy and democracy[43] that preceded the age of kings, as either unimportant or non-existent.

The warning we are getting at is that when we try to project a main line of history, or an 'end of history' analysis, we will inevitably gloss over significant variations in the central trend, and we may ignore offshoot trends that become central later on. Who could have predicted in 500 BC that Greece, rather than Persia, would become the cultural model for the middle eastern world?

THE AGE OF HIGH-TECHNOLOGY INDUSTRIAL CAPITALISM AND LEGAL-REPRESENTATIVE DEMOCRACY

We could be entering a new historical era. There is little doubt that high-technology capitalism, in one form or another, will be the mode of production and that parliamentary democracy will be the form of government, in one or another of its variations. Legal authority[44] will be the legitimation system, representative government with free elections will be the form of government, and the industrial system will be capitalist.

Yet, even if this is so, many questions will remain – questions whose answers point up the possibility of a different kind of historical era.

For instance, there is the bureaucratic question: will government officials, as in Japan, become more powerful than the legislature? This is the 'new mandarin' problem which Chomsky[45] has raised, and which was raised earlier and in more detail by Mills in the *Power Elite*,[46] and earlier yet by Weber.[47]

In its extreme formulation, this vision projects the view of a world more like ancient Egypt than ancient Greece. It envisions a world run by 'soulless' technocrats and occupied by a people turned inward toward personal pleasure. Will we have a world run by MITI-style mandarins – a world overflowing with high-technology toys, accompanied by a decline in political participation and freedom? Or, will we turn away from the abundance, and limit such bureaucratic control?

'. . . and the children of Israel fled from slavery and turned their backs on the fleshpots of Egypt and its storehouses of food and goods and material things. . . .'[48]

Then there is the democratic-socialist problem, raised by its critics, such as Hayek[49] and Popper.[50] Will welfare-state socialism, as in Sweden, and now the entire European Community, lead to a paternalistic planning system, not unlike that of the Japanese system (if less Spartan and more oriented toward social support programmes)? This is the 'mandarin' problem again, but emerging from a different ideological origin.

In a completely different vein, there is still the old Marxist question about whether in a capitalist economic system – any capitalist economic system – wealth will not be selfishly accumulated and controlled by a few, leading to oligarchical–plutocratic over-influence in the legal-representative system.

This problem – oligarchy of the rich – is still as much of a problem for capitalist societies now – as Kevin Phillips has shown – as it was during the 1890s and 1920s.[51] In the United States, has not the Senate become a 'millionaires' club', and are not campaign financing and the control of lobbyists major political issues? Did we not live through the worst stock market and banking scandals in our history in the 1980s?

Therefore, even though the American system is deeply grounded in law, anti-bureaucratic and democratic to the core, and even though we rightfully hold out our Bill of Rights and 'rags to riches' individualism as gifts to the world to emulate – still the old spectre of plutocracy haunts this variation of main-line historical development.

Finally, and significantly, the mass media and the whole new communications process have just begun to emerge as central institutions. We watch the world through CNN, and talk to each other through fax and e-Mail and the Internet and (soon) television-telephones. The 'global village' is happening, but just what this will mean politically, socially and economically, we do not yet know.

Certainly, transnational corporations become possible, even logical. But, the national interest is not always served by transnational economic units. Look at the struggle now occurring between the United States and Japan, even though Toyota and General Motors have joint ventures.[52]

Will there be a decline in nationalism, as Robert Reich seems to be suggesting in his 'who is us' thesis?[53] Or will there be a rise in nationalism as stated by Thurow[54] as the new economic units spread their benefits unevenly? Look at all the Japan-bashing[55] that is going on right now in the USA, and the terrifying rise in ethnic conflict that has accompanied the break-up of the Soviet Union.[56]

Will the mass-mediated global culture create greater understanding between people and engender a long-term homogeneity, or will the media just heighten ethnic awareness and ethnic conflict? Will the mass media and new communications networks foster free speech and expand the democratic potential of mass societies, or will the media standardize perceptions and narrow the vision of viewers?

For any social scientist attempting to understand our near-future world, the communications revolution must be analysed. Yet, since these new systems are not yet developed, speculation as to their impact is risky indeed.

CONCLUSIONS

Fukayama – standing Marx on his head – has written of the 'end of history'. There is no end to history, but there is usually a main line of historical development. Following Marxian causality, we shall argue that from around 1965 onward, a new mode of production has evolved that has altered the course of history once again. This new mode of production is high-technology industrial capitalism, wherein computers oversee robots which tend machines attended by human workers, which produce everything from advanced electronic products to high-technology sneakers. The whole production system is linked up with communications networks that are world-wide.

This new mode of production has engendered new classes – technocratic, bureaucratic, service, and white-collar office workers – who will have a significant impact upon the socio-political structure of near-future societies.

The new mode of production has also shifted the locations of the majority of the industrial working class, from the advanced industrial nations to the developing nations. And, unfortunately, because of the level of technical proficiency demanded of the new white-collar and blue-collar work force, an underclass has also emerged with the new mode of production. This underclass could be a temporary class, or more long-lasting. But, its political impact has already been profound in terms of crime, social unrest and ethnic conflict.

The question most central for us is that of the survival of democracy on this new capitalist base. Will the new middle class become the 'carrying class' for a modern form of democracy utilizing the sophisticated communications technology to expand the democratic potential of mass society? Or will democracy decline under the weight of the managerial and technocratic strata so essential to the functioning of the modern economic and political enterprises?

Finally, technological totalitarianism – 1984 – is also part of the historical potential.

Let us conclude this chapter on a theoretical note: the causality of historical change is not unilinear. That is, the economic and technological systems – or material causalities – are not the sole forces of history. Political, ideological, and cultural forces may act independently to alter society at large. The Japanese case, for instance, was one wherein political decisions and cultural orientations actively shaped the economic system that emerged,[57] though the establishment of the industrial-capitalist economic system was the basic motivational goal of Japanese historical development. But this goal itself was motivationally rooted in military and political desires, rather than economic ones.

In our analysis of the near-future capitalist democratic societies, therefore, we shall attempt to present a balanced view of causality and a complete picture of social organizations.

We shall begin with an analysis of the new mode of production, however, because in our era, from the 1960s to the present, the economic system has undergone a rapid and remarkable transformation (as astonishing as that of the first industrial revolution), and it seems to be leading the other institutional spheres for the time being. Japan, after all, was not catapulted into the spotlight because of its political system or cultural creativity, but because of its remarkable high-technology productivity.

So, without falling into the Marxist unilear, teleological trap, let us first look at the high-technology industrial capitalist mode of production, and its socio-political impact on the emerging historical world.

Part I

Legal-Representative Democracy on a High-Technology Industrial Capitalist Base

1 High-Technology Industrial Capitalism as a New Mode of Production

HIGH-TECHNOLOGY INDUSTRIAL CAPITALISM

We are theorizing that industrial capitalism has evolved to a new level of productive capacity, and that the structure of the forces of production are so different from the old-fashioned smoke-stack factory system as to have engendered a whole new set of social relations. In Marxian terms, a new mode of production has evolved (but not the one that Marx had envisioned, of course).

Economic theorists from Berle and Means[1] and Burnham[2] to Galbraith[3] have been describing the dramatic changes that have taken place in the structure of the new economy. By the 1980s it was clear that something very different had emerged in terms of the economic system and the professional roles connected with it. Yet no one had given the new mode of production a name.

Finally, Daniel Bell,[4] realizing that we no longer lived with the kind of industrial capitalist economy described and analysed by the classical economists (from Smith to Marx) popularized the concept of a 'post-industrial' society. The problem with this characterization – which has caught on – is that the new society is still *industrial*. The old-fashioned factories may be disappearing, but their place is being taken by computerized, robotized, automated 'factories', in the developed nations, and by high-tech 'sweat shops' in the Third World.

The high-tech factories, in the developed and developing nations, still produce the goods everyone either needs or desires. The majority of the work force may no longer be focused directly on this productive process, but how would a 'service' sector of the economy be possible without it? The economy is still centred, after all, on the production of food, clothing, housing, consumer goods and military weaponry. Education, health care, child care and other services are, of course, critical and growing, as Bell has correctly pointed out. But these services are either infrastructural to the new high-tech economy, or take for granted the high-technology

17

ability to produce a *superabundance* of all life-sustaining, or life-style enticing, consumer goods.

Thus, there are a number of characteristics of this new economic system that need analysis before we can really understand it.

First, it must be established that the new mode of production is *industrial*. That is, it is not 'post-industrial' except in its time sequence. Industrial products and industrial production are still the basic components of the economic system. In fact, high-technology products have swept us away in a tidal wave of world consumerism. The new economy has a vast service and bureaucratic sector, as Bell and Burnham have emphasized, but the desire for high-technology products motivates the system at its core.

Second, the new industrial economy utilizes *advanced technology* to produce consumer goods, and the consumer goods themselves incorporate the technological advances. Thus, both the 'factory' and its products have been altered by the technological revolution.

The factory system pioneered by Henry Ford – with an assembly line and workers whose tasks were simplified and standardized – has been replaced. The new factory system requires technocratic workers coordinated by a managerial-administrative hierarchy. And, even the line workers on the shop floor must be able to provide feedback to the technocrats and managers. 'Fordism' is now passé. Involved shop-floor workers, with technocratic, managerial and financial teams supervising them, have replaced the factory assembly line.

Be advised, however, that a very large working class is still required in this high-tech industrial process, for the assembly of products, packing, shipping and other 'low-tech' industrial necessities. This working class should not be lost sight of, though the technocratic, managerial and financial strata are emerging as new and important classes.

Third, the new mode of production is *capitalist*. Up until the communist collapse in Russia and Eastern Europe, most of the world believed that an advanced industrial economy could be non-capitalist. In fact, many intellectuals – socialist[5] and non-socialist[6] – believed that capitalism would decline and the world's industrial economies would become increasingly socialistic and planned.

However, it now has become clear – as Keynes[7] and some of the Swedish economists[8] had maintained all along – that the capitalist segment of the modern economic system is critical. Efficiency, pricing, entrepreneurial creativity, profit motivation, the competitive market, banking, loans and interest calculation, and yes, even the stock market for capital accumulation and investment (scandal-ridden though it has become) – all of these capitalist processes and institutions are a necessary and integral part of the new-high technology industrial system.

Let us look more closely at each of the three phenomena we have focused on.

THE HIGH-TECHNOLOGY ECONOMIC SYSTEM AS BASICALLY INDUSTRIAL

Daniel Bell, as mentioned, in his *Coming of Post-Industrial Society*[9] has written a fascinating analysis of the emerging world. But, as we have pointed out, the title of the book – and its central thesis – are highly misleading. The new economy is, of course, no longer the economy of 'smoke-stack' industries, of heavy, clumsy machines tended by thousands of semi-skilled assembly line workers. However, the new economy is still an industrial economy. The machines are super-high-technology electronic machines, and they are tended by computers and robots, which in turn, are tended by technocratic and managerial white-collar workers, with blue-collar workers still essential at the shop-floor level.

But, what is it that is produced? What is it that everyone wants? The answer is clear: industrial products. No, we do not want Model-T Fords, rather, we want BMWS with in-board computers, 24 valves, anti-lock disc brakes, Mcpherson front suspension, all-wheel drive, electronic fuel injection, electronic overdrive, turbo-charged acceleration, automatic transmission, digital dashboard, disc stereo, dolby FM, and so on.

We still want a car, don't we? And that car has to be produced in a mass production factory. It does take scientists, engineers and other highly educated and trained technocrats to invent, design and produce such a car, but it is the car that is at the centre of the process.

Bell emphasizes the importance of the 'service' sector of the new economy. And, the service sector is truly important. There is a whole new world of health, educational, social work and other services – both governmental and private – which we cannot live without in the modern world.

It is highly significant, however, that in China, Russia and Eastern Europe, where health services, the educational system, and government services in general, were well developed in relation to industrial production, the absence of consumer abundance engendered such discontent, that along with the genuine desire for political freedom, it produced a revolution against the communist system.[10]

It is not that the service sector of the economy is not growing, for it surely is – and we will carefully analyse the new middle-class strata being engendered by it. It is that the industrial productive sector of the economy is still the base upon which all else is built. After all, Japan, with the smallest service sector of the economy, comparatively to the European Union and the USA, has become the legendary leader of the economic world because of its inundation of the world's markets with stellar, sophisticated, high-technology products. Who asks about Japan's hospitals? Are Japan's schools at the cutting edge? Sony, Honda, Toyota and the other high-technology giants of Japan define the success of the new mode of production.

Is it really the service sector that characterizes the new economy, or the micro-chip circuit, the transistor, the computer, the fax machine, the copy machine, high-definition television, the VCR, super cars, multi-function quartz watches, world cable and satellite communications networks, and even specialized high performance sneakers, specifically developed for running, walking, biking, tennis, basketball, aerobics, and cross-training?

One more factor must be emphasized in terms of the industrial nature of the new mode of production. Though white-collar and service workers are replacing blue-collar factory labour in the advanced nations, the high-technology firms have increasingly been moving their operations to the Third World[11] – especially the Pacific Rim, but also Latin America. Within these less developed nations, mass industrial factory production has expanded dramatically. Searching for cheap labour, American, Japanese, and European firms have moved their plants to Taiwan, Singapore, Korea, Thailand, Indonesia and a number of Pacific Rim islands (and now the free-trade zones in China), along with Mexico, Brazil, and other cheap-labour locales.

Thus, industrial production – of high-technology products, like VCRs, and of old-fashioned industrial products, like clothing and shoes – has not declined at all, but rather it has moved. The design of the products may come from the advanced nations, and may include sophisticated electronic devices, but the products are still produced on an assembly line, utilizing cheap, semi-skilled labour. Automation may, in the long run, eliminate the need for an industrial labouring class, but for now it has only shifted its location.

Still, the automation has major consequences for the class balance within the advanced industrial nations, since the working class declines in numbers, while the white-collar classes increase. And the shift of production locale has had an equally monumental impact on those third-world nations now integrated into the high-technology, industrial capitalist network of transnational production.

'Globalization' is the term gaining currency. And, this globalization is not only altering the class structure of the nations involved in it, but also generating a new world culture, centred around technological production and trade, communications linkages, and mass-media news and entertainment viewing. We shall have more to say about the global economy and the world culture in terms of their positive or negative effect on political democracy.

THE NEW MODE OF PRODUCTION IS CAPITALIST

Veblen[12] made a distinction between the industrial portion of the economy and the financial end of the economy. He believed that in the United States, during the early twentieth century, we were paying too much attention to finance and not enough to industrial production. His solution was for the 'engineers' to take more control of the productive sector, while the financiers played a lesser role.

The German economy of the 1880s to the 1920s period had placed greater emphasis on engineering than on finance – as Veblen pointed out.[13] However, and Veblen was fully aware of this, the business end of the German economy was still central, and in fact, developed a strategy for catching up with Britain and the United States based on a manipulation of the market principles. This latter was enunciated by the little-known economic theorist, Frederich List.[14]

It is with the communist economies that business principles and market economics were abandoned. Lenin and his clique believed that a society could create a successful industrial system without reliance on the capitalist market institutions. When Lenin saw that his ideology might be incorrect, he helped establish a 'new economic policy', which began to incorporate market principles, along with the planning process of the communist party.

NEP might have evolved toward a system something like that of Imperial Germany in the nineteenth century. We shall never know. For, with Lenin's untimely death, Stalin not only abandoned the NEP market policies, but banished them from the Soviet realm. With Germany menacing the Soviet Union in the West and Japan invading the Pacific East, Stalin developed the military-style command economy that became the model for all of the communist nations that emerged after the Second World War.

Even then, because of Russia's and China's successful military buildup, economic theorists, and most scholars and statesmen believed that a modern industrial economy could exist without capitalist institutions and processes (and most Marxist intellectuals believed that democracy could exist without the Enlightenment principles of power limitation and rule by constitutional law).[15]

The collapse of the communist system, however, has made it clear that capitalist institutions are necessary for the proper functioning of the industrial economy. We shall come back to Veblen's distinction later, for it is now also obvious that the industrial sector is dependent upon scientists, engineers and other technocrats for its proper functioning. However, the centrality of market principles is also now fully clear – as is the centrality of the Enlightenment political principles – including legal authority and the limitation of the power of the state; but we shall discuss these processes later.

Returning to the economic dynamics, Max Weber[16] had insisted before World War I that capitalist principles facilitated both efficiency and technological creativity in the industrial economy, and that bureaucratic state socialism would inhibit industrial growth. He was absolutely correct. For, though Adam Smith never really related his classical economic dynamics to his 'pin factory',[17] the capitalist control of industrial factory production was considered a prerequisite for economic success.

As we approach the end of the twentieth century, it has become clear that the capitalist principles of supply and demand, market pricing, profit motivation, interest-bearing loans, business competition, free trade and tariff trade strategies, the stock market and the banks, and so on, are absolutely necessary for the efficient functioning of the industrial productive system. Thus, the new mode of production is, and will continue to be, capitalist in its structure and culture. Market principles, though often modified, will direct business and industrial activities.

It must be made clear, in this vein, that the economies of the Scandinavian countries, though socialist in terms of their social service and safety-net programmes, are fully capitalist at their core, and becoming swept up in EU style ideology – with more emphasis on productive efficiency and cutting-edge technology and less on safety-net and service programmes.[18]

It should also be mentioned that though the Japanese actively intervene in, and even contravene, market processes, the Japanese economy is rooted firmly in market dynamics. The Japanese couldn't intervene so effectively, if they were not experts in the capitalist principles![19]

Now, as we shall make clear, the Japanese, German and American systems represent three different models of capitalist economic organization. And the differences may be important in terms of their influence on modern politics. This we shall soon analyse. Here we wish to emphasize the capitalist nature of all three models.

The German and Japanese systems include a good deal of government planning. And, government–corporate labour cooperation makes these systems different from the classical Anglo-Saxon model. Given the remarkable performance level of the Japanese economy, and the very good performance level of the German economy, future models of production may evolve further toward some modified version of capitalism, including state intervention and 'corporatist' structures.[20] The United States, however, reflecting a more classical arrangement of corporate and government relations, and a more confrontational system of labour–business relations, has continued to lead the world in entrepreneurial creativity and technological breakthroughs. On the other hand, Germany and Japan are beginning to become leaders in research and development and to run even with the US in this regard.[21]

And it must be pointed out, lest the reader get some naive or romantic notion about American capitalism mirroring some Smithian utopia, that neither Adam Smith, nor Milton Friedman[22] can make the invisible hand of the market fully regulate the high-tech giant corporations and conglomerates that now dominate the modern economy. These firms will monopolize (oligopolize) markets,[23] stimulate demand through massive advertising campaigns,[24] control supplies or suppliers,[25] and fix prices and rig stock market fluctuations. This is not pure market dynamics.

Further, between 1940 and 1988, the United States government engaged in 'military Keynesian' investment on an unprecedented scale. During the Reagan administration alone, the United States government invested nearly 3 trillion dollars in defence-related production, research and development. This kind of government investment characterized the American economy from the Second World War to the Gulf War.

Finally, the recent experience with de-control and de-regulation has been disastrous. The collapse of the banks and stock market should serve as a warning that the days of unbridled capitalism are over.[26]

Having asserted this, we wish to make the opposite point: that capitalism, though in some sort of modified form, will be an integral part of the high-technology economy of the future, because capitalism motivates productivity, creativity and efficiency.

It is important that the new economy will be capitalist in its orientation, because it means that the capitalist institutions and classes will continue to exist, and, as we have shown previously in *For Democracy: The Noble Character and Tragic Flaws of the Middle Class*,[27] and as we shall show again, these institutions and classes have been structurally and ideologically linked with legal-democracy.

As I described in detail in the *For Democracy*, contract law originally undergirded constitutional law, the separation of the economy from state control led to the conception of the power-limited state, free enterprise and free citizenship go hand in hand, and even the rational-scientific world-view emerged as part of the rational calculation of the commercial-capitalist cultures of Greece and Renaissance Europe. The rational-minded free citizen recognizes no kings or hereditary lords, and demands self-rule and the protection of law.

Thus, legal-representative democracy has been undergirded by capitalism – first, trade capitalism, and then industrial capitalism. The question for us now is: will high-technology industrial capitalism continue to undergird democracy?

THE TECHNOCRATIC AND BUREAUCRATIC ALTERATIONS IN THE STRUCTURE OF THE MODERN ECONOMY

Veblen, though overemphasizing the role of the 'engineers' in the modern productive process, was absolutely correct in focusing on the new, central role of technocratic-scientific experts in the productive process.

These technocrats have become central, not only to innovations in the productive process, but also to the invention of the products themselves. No longer do we find the 'eccentric inventor' alone in his basement, creating the light bulb. Rather, teams of technocrats – supported by huge research and development budgets, connected to the corporations themselves, or government or university facilities – invent, improve and develop the high-tech products that we crave, and then also improve and develop the processes by which they are made. Therefore, the modern economy must have a technological component separate from the business component, exactly as Veblen had described it.

However, production has become a very complex process. Since the high-technology firm is capitalist, it requires a large-scale business sector. And, since it is industrial, it requires a labouring force and factory set-up. And, since it relies on high technology, it requires research and development facilities with teams of technocrats. All these complex divisions of the firm must be coordinated and directed by a large-scale bureaucratic-administrative-managerial organization. Given the transnational or multi-national nature of these managerial operations, the administrative end of the modern corporation can become complex indeed.

Because the administrative functions of the modern high-tech corporation have been taken over by managers, these corporate executives have grown in numbers and power within the capitalist-industrial system. This, of course, is the Weberian analysis of the giant corporations which we shall discuss later on in this treatise.[28]

Suffice it to say here, the managers have emerged as a new power class in modern society. As such, the base of their power must be analysed, as well as those countervailing powers (such as those of the business and labouring sectors) which may oppose them.

Furthermore, the cultural affinities which the managers carry, as a new potential 'ruling class', in the Marxian sense, must be carefully examined. Will they be a new 'mandarin' upper class?[29] Do they carry any democratic elective affinities? Will this managerial class act as a class for themselves, or continue under the control of the finance-based capitalist stratum?

As the situation now exists, it seems that the corporate manager must somehow combine the skills of business, engineering and management in order to succeed. Since this new structure is bureaucratic, capitalist and technocratic, the nuances of this complex corporate organization are still being studied world wide – along, of course, with the two other unanswered questions: the degree of government involvement in the production process, and the role of labour as cooperative partner or adversary.

Another remarkable capitalist phenomenon has taken a quantum leap beyond its previous functioning. That is, the high-technology, computer-linked information networks have now been harnessed to the world's financial markets. The application of computer technology to the world's capital markets has created a new system of electronic, global finance, and a new class which could have a powerful effect on the economic and political systems emerging. This 'new class' is not the bureaucratic-managerial class of Burnham,[30] Berle and Means,[31] or Djilas.[32] This new class is a stratum of financial speculators, 'betting' on everything from US Treasury bonds to yen, but hedging their bets with mathematical formulas built into the computer software of the global financial system.

Will this new high-tech financial stratum dominate the managers? They certainly have the managers frightened in the USA. But, as yet, the financiers are not strong in Japan or Germany. If the financiers dominate in a nation, will industrial production decline or move elsewhere? If financiers make and lose billions on a daily basis, how does this affect the world and national economies? And finally, with the gold standard yielding to the new high-tech 'megabyte' standard, what will happen to the world's investment process – has the stock market become divorced from the industrial economy?

THE COMPUTERIZED, GLOBALIZED FINANCIAL SYSTEM

Not only are manufactured products and manufacturing techniques now highly technologized, but the financial markets have become technologized

as well. Today's capital markets are utilizing computers and highly sophisticated software developed by Nobel Prize-winning mathematicians and
economists.

This computerization of the world's financial markets has had remarkable socio-economic consequences. It has further separated 'ownership'
from control of the modern corporation – especially in the United States
and Britain, and it has created new opportunities for arbitrage-trading.
This latter has engendered a world-wide speculation so vast that hundreds
of millions of dollars sometimes change hands within seconds on the
world's financial markets.

On the positive side, the telecommunications–computer revolution has
created a world capital market. This has occurred partly because of institutional changes, such as the elimination of the gold standard in the 1970s
and the creation of 'floating currencies', along with the deregulation of the
financial markets which occurred in the 1980s. However, the change has
occurred more centrally because of the application of computer technology to the financial markets.[33]

The computerization and global telecommunications connection has
encouraged 'global sourcing', or the setting up of high-technology manufacturing facilities anywhere in the world. The world capital market has
made it easier for poor countries to export to rich countries and for rich
countries to establish production facilities in poor countries. Wealthy countries will still accumulate more capital, but their 'savings' will flow into
a world capital market where they will be allocated to the regions generating the highest returns.[34]

Effectively, everyone now has access to the same world capital market.
More equal access to capital has reduced the edge that being born in a rich
country used to give. This explains why countries like Korea and Singapore
could gain the capital to expand their industries, and why they are able to
sell abroad so easily.

Thus, the global computerized financial market has had positive effects
for nations such as those of the Pacific Rim. These nations are now gaining plants on their soil. And, if most of the jobs they generate are low-
paying industrial jobs, managerial, technocratic and white-collar jobs are
also expanding, along with the industrial base of these nations.

Of course, the effect on the 'First World' countries of this international
'sourcing' and global financial access has been to reduce their industrial
base and shift the economy toward the service and the financial sectors.

What is this new computerized, globalized financial system, and how
did it arise? How will it affect industrial-capitalist productivity? Will it
affect legal-representative democracy?

MONEY: FROM THE SHEKEL TO THE MEGABYTE

Money, as a unit of exchange with a specified value, has been utilized by human societies since the period in history when craft goods and food stuffs were produced in enough of an excess to make trade possible. This phenomenon occurred at different times in different areas of the world. Since the direct exchange of one kind of goods for another was ineffective once a wide variety of goods became available and distances of trade lengthened, barter gave way to money-exchanges.

In most societies of the world, as they moved to the level of open-field agriculture and specialized craft industry, metal became the medium of exchange. Metal was highly valued in itself, and it was light enough to carry around in bags, yet it was heavy enough to seem impressive. More importantly, metal was long-lasting, and, the seal of the king or the high priest could be imprinted upon it, giving it the added weight of sacredness and power.

Since Sumeria was probably the oldest civilization (rivalling Egypt for this claim), we should not be surprised to find that money was probably first invented in the temples of Sumer about 5000 years ago. It was a store of value, a unit of account for the temple priests. Sumerian money was valued as a lump of bronze equal to a Sumerian bushel of barley, or 'shay' – hence the name of the first coin, the 'shekel'.[35]

Since bronze and iron became important as military and industrial products, there was a tendency to fashion coins from silver and gold. These metals were beautiful, but too soft for weapons or tools. Thus, from Babylon to Hellas to Rome, silver and gold coins came to be the money of choice.

If we skip the centuries, and arrive at the trading world that emerged from Renaissance Europe, silver and gold became the European monetary standard. The gold standard emerged as the eventual evaluator of the world's money. From the Spanish conquistadors to the American industrialists, gold became the stabilizing valuator underlying all national currencies, whether metal or paper.

With the decline of the Spanish, the British, with their burgeoning empire and their industrial revolution, carried the gold standard to the world. Gold and the British pound sterling brought stability to the world's monetary system right through to the First World War.

After the Second World War, the world economy was governed by what was called the Bretton Woods Agreement. The main feature of this agreement was that all of the world's major trading currencies were rigidly linked to the dollar in a system of fixed exchange rates. The dollar, at the centre of the system, had its value anchored to gold, at the official rate of

$35 an ounce. Because the dollar was convertible to gold and all other currencies were convertible to dollars at a fixed rate, the world was effectively on the gold standard.[36]

Just as the British economy had been devastated by First World War debt, the American economy was disrupted by the Vietnam War debt. Billions of dollars had been expended on the Vietnam War between 1965 and 1973. During this same period, German and Japanese industry reached a competitive advantage with American industry.

President Nixon, running for re-election in 1972, was worried that the economy could be his downfall. Therefore, in a televised speech, Nixon announced a presidential order freezing wages and prices for 90 days, limiting strikes, placing a 10 per cent surcharge on foreign goods, and he said that he had 'closed the gold window'.[37]

On 15 August 1971, money – in the old sense – was repealed. Nixon transformed it into something totally new – currency without any underlying value whatsoever, and without any limitations on the government's ability to create it. Money was to become a 'computer abstraction'.[38]

THE NEW ELECTRONIC-COMPUTERIZED-GLOBAL FINANCIAL SYSTEM

Money has been transmogrified . . . it is now a system. Money is a network that comprises hundreds of thousands of computers of every type, wired together. The network of money includes all the world's markets – stock, bond, futures, currency, interest rates, options, and so on.[39]

The network is juxtaposed by computers that chart investment-risk using the Nobel Prize-winning formulas developed by Harry Markowitz and Merton Miller of the University of Chicago, and other esoteric and advanced mathematical equations and programmes.

Every day in New York, more than 1.9 trillion dollars electronically changes hands at nearly the speed of light. These dollars appear as momentary flashes on a screen. Sums of similar magnitude pass through the fibre-optic networks of Tokyo, London, Frankfurt, Chicago and Hong Kong.

The financial system, using money that has been liberated from gold, each day conducts transactions hundreds of times larger than those in the so-called real economy – the part of humanity's endeavours where goods are produced and services sold.[40]

The computerized, globalized financial system allows traders to conduct rapid electronic transactions, with no necessary knowledge of the producer-firms involved. This ascending of trading means that finance has shifted its orientation from one of investing to one of transacting. The information era has moved finance from its age-old static – even passive – framework of buying assets or making investments and then holding them, into something new: making money by harvesting minute shifts in value across various exchanges.[41]

Traders are no longer expected to know about a company's history or even its management. If a stock fits a desired mathematical formula for price, volatility and dividends, it is bought, no matter what the underlying company makes or who manages it.[42]

THE DIVERSIFIED PORTFOLIO AND THE END OF OWNERSHIP

By constructing portfolios that are no longer based on individual stocks and the companies they represent, a process of divorcement began in the financial community, with tremendous repercussions. Stock shares were essentially separated from their industrial ownership base.[43] The financier no longer owns a share of industrial stock, such as General Motors, but rather owns a share in a mathematical computerized system for financial speculation. The contemporary investor is not an investor–owner in the capitalist industrial system, but rather an electronic gambler, minimizing risk through high-technology strategies, and completely divorced from the actual productive system. In fact, the company matters less than the stock, for switching in and out of risk groups – under the guidance of programmed equations – has become the newest investment fashion.[44]

Beyond Berle and Means

With diversified portfolios, stocks are no longer 'owned'. However, if stocks are not owned, who owns the corporations? No one. They are bought up, bought off, traded, re-traded, expanded, conglomerated, broken apart, reassembled, transnationalized, and often, lost completely.

All of this is done by financial cliques, financial firms, managerial cliques, or foreign investors – none of whom are interested in the industrial productive process or the technological innovations inherent in it.

Have the American managers lost control in Berle and Means' sense? Yes. They are forced into short-term profit strategies, or forced to buy up their own stock, and forced away from R&D and long-term investment. They hold their 'golden parachutes', ready to bail out whenever the firm is ready to crash.

Veblen's nightmare has become real – at least in the USA. The 'captains of finance' have overridden not only the engineers, but the managers as well.

But economic history does not begin and end in the 'Anglo-Saxon' nations. In fact, it may be advancing in Europe and Japan, where Veblen's theoretical prescriptions for the industrial-productive sector of the economy has not been lost. Thus, while the 'megabyte' capital system is developed and pursued in America, the high-tech industrial system is being developed and pursued by our competitors.

In the meantime, the global capital system with its ever-improving computer technology, is developing with a dynamic of its own, at once connected to the industrial system through the world of global investment, international trade and the world currencies market, and at the same time, divorced from the industrial base in terms of the 'producer' necessities of long-term investment, R&D, and smaller immediate profit-taking.

The whole process of world-wide capital investment in industrial units has become divided between the world of electronic computerized globalized finance, and the world of 'industrial rationalization'.

The European Union and Japan have been emphasizing industrial rationalization, while the Americans have been putting their efforts into world-wide financial institutions.

CONCLUSIONS

Having described the outlines of the new high-technology industrial capitalist system, let us now look at the three different models of high-technology industrial capitalism emerging in the United States, Europe and Japan. Will they remain different or converge? Do the different models have a differing impact on the legal-democratic political systems they undergird?

We must look carefully, for in the European and Japanese systems, the bureaucratic managers have more power then the financiers, while the reverse is true in the United States. And, while the European and Japanese systems have become relatively egalitarian in terms of wealth distribution, the American system has skewed toward a growing inequality.

2 Three Models of High-Technology Industrial Capitalism

INTRODUCTION

It was the British who invented the industrial form of capitalism – its institutions, its processes, and its theory. Trade capitalism had been around since ancient times, and it had been modernized in Europe during the Renaissance and Reformation. Industrial capitalism was something new, as Smith, Bentham, Ricardo, Mill and Marx understood.

Why did the industrial revolution occur in Britain during the eighteenth and nineteenth centuries? Ask Weber.[1] Ask Tawney.[2] Whatever your theory, the fact remains: industrial capitalism developed first in Britain.

Both the British economic system and British economic theory exhibited a combination of industrial and capitalist principles. Thus, Adam Smith waxed eloquent about the pin-factory, citing its industrial rationalization as superior to that of craft production. He also soared into intellectual history with his theory of the competitive market and its self-regulating processes pertaining to price, profit and production.[3]

It is important to note that the industrial rationalization portion of British economic theory was downplayed, after Smith, while the market dynamics were intensely studied and theorized upon. This split is important. For the industrial portion of economic theory would be neglected in Britain in favour of the market portion. The development of market theory – from Smith to Marshall to Keynes – would enshrine Britain as the fount of classical capitalist economics.

The industrial rationalization portion of the theory (and practice) would be developed later, in the United States, where, from Veblen to Taylor to Ford, the efficiency and productivity of the industrial system itself would be improved and analysed. As we shall show, the Germans and then the Japanese would focus their theory and practice on industrial rationalization, whereas the British and post-Second World War Americans would focus on market theory.

32   *The New Middle Class and Democracy*

THE BRITISH AND AMERICAN MODEL: BUSINESS, GLOBAL FINANCE, AND FREE MARKET ECONOMICS

The British and the Americans were creating an industrial revolution. Therefore, industrial factories were built, and then improved, in both nations. It was the industrialization – the machines that produced the goods, and the machines that produced new machines (such as trains and cars) – that generated a new source of power. Machines could be used both for peace and for war, and therefore, suddenly, the balance of power in the world tipped toward the emerging industrial nations. Thus, England and then the USA became world powers, replacing Spain, Germany, France, Russia and China.

It was the industrialization which caught the attention of the world. And, it was the factories and industrial products – and the power these bestowed – which other nations wished to possess.

Even though it was the industrial portion of industrial capitalism which the world's nations desired, it must be understood that within Britain and the United States, the industrialization was carried out by an independent class of capitalists, who thought of themselves as 'businessmen' first and foremost. They thought and acted in terms of business expansion and rising profits. They were 'captains of industry', as Veblen called them, and they often combined the talents of inventor and industrial innovator with the skill of business entrepreneur. Yet, they remained businessmen first, employing and improving technology as rapidly as they could because it produced a profit for them. They would employ 'engineers' and 'efficiency experts', as long as this led to business success.

The capitalist portion of the industrial capitalist economy was the motivational engine that drove the industrial machine.

As Veblen[4] pointed out, in both England and the USA, the 'captains of industry' became 'captains of finance' by the early twentieth century. Thus, though Great Britain led the world in industrial production, and the USA then took over the lead, the capitalist sector of the economy has always generated greater veneration than the industrial sector of the economy. Why should this be so?

The Obstructive Kingly–Bureaucratic State vs Free Capitalist Economic Development

From Smith, Bentham and other classical capitalist economists, the world received the theory of the 'free market' – the unfettered, self-regulating market, the 'invisible hand' of the market creating the perfect economic equilibrium, for the nation and for the world.

The other side of this equation of economic perfection was the obstructive state – the regressive, bureaucratic, inhibiting state, bent on control of the economic forces and on their gross misuse and mismanagement.

The notion of the obstructive, backward-thinking, meddling state grew out of the political realities from which the industrial capitalist revolution emerged.

In England, France, Germany, Austria and Spain, the monarchical governments were attempting to actively intervene in the economic processes which had been developed independently by the businessmen and craftsmen of the free trading cities of the late Renaissance. These cities – from the Italian to the Hanseatic and all across Europe as far north as Bergen and as far east as Novgorod – had developed a successful trade-capitalist economy which was on the verge of expanding world-wide.

However, the kingships were beginning to drain them dry of their hard-earned capital, and threaten their independent, dynamic economic activity. The policies of the monarchies were disastrous economically: looking to military expansion and luxury living, the monarchies were extorting all the money they could from the businessmen, artisans and peasants of their realms.

It was within this context that the classical capitalist economists wrote. The free market became a political, as well as an economic issue, and the separation of the state from economic activity became a creed rather than merely a theory. 'That state which governed best would be that state which governed least.' The minimal state – the state that did not interfere with business practices – was the ideal state. And, the self-regulating market and the minimal power-limited state would form the foundation, not only for dynamic capitalist economics, but also for legal-representative democracy.

The revolution by the citizens of the free cities against the monarchies of Europe was a democratic as well as capitalist revolution. (It was also a religious revolution, and this we shall shortly discuss.)

Since the kingly-bureaucratic state was both the enemy of the free market and the enemy of the free citizen, the ideal of the self-regulating market and the power-limited minimalist state generated democratic legitimacy as well as capitalist legitimacy. The democratic sanctification added an extra halo to the free market concept. For, if the market were truly self-regulating and would engender dynamic economic growth, then a greater political freedom would be possible since the state could be less powerful internally (a strong external military force was allowable within Smith's utopian conception).

If the democratic legitimation were not enough, the whole free-market, free-citizen conception was more mightily legitimized by its puritan–protestant religious orientation to the world.

Puritanism and the Legitimation of Work, Money-Accumulation, and Individualism

The link between protestantism, capitalism and democracy is well known and well argued.[5]

The points we wish to make here are these: Luther's debate over 'good works' for the church vs faith were misinterpreted by the hard-working folk of the German cities. The 'vulgarized' Lutheran doctrine that emerged was that 'hard work in one's calling gained one heavenly grace'. Therefore, where protestantism was established, hard work was enshrined, and, hard work was necessary at the industrial factories.

Second, Calvin's doctrine of 'inner-worldly asceticism' – or a rejection of the pleasures of the flesh – combined with his doctrine of predestination, also gave rise to a 'vulgarized' version. From the artisan trading cities of Switzerland and Holland and West Germany, the peculiar interpretation emerged that if one worked hard in one's calling, and if one made money, and if one used that money to make more money, the resulting accumulation of money-wealth was a 'sign' of heavenly grace. One could not use the money to luxuriate the church or one's own household, because of the puritan rejection of this-worldly pleasures and luxuries. Therefore, one had to re-invest the money, in order to make more money. This, of course, is Weber's theory of the 'spirit of capitalism' emerging from the vulgarized doctrines of the puritan–protestant reformation.

Thirdly, the protestant insistence that each individual had the right to interpret scripture for him- or herself, and that no church hierarchy or civil authority could interpret scripture for the person, led to the ideal of 'individualism'. The individual as minister, the individual as citizen, and the individual as entrepreneur.

Why mention puritan protestantism? Because in Britain and the United States, the economic and the political system – capitalism and democracy – gained a godly aura. The businessmen were 'god's stewards on earth', giving every worker a money-wage and starting them on their road to grace. The democratic constitution and the laws were given by the grace of God, and the citizen's rights were inalienable natural rights, god-given.

To sum up then: because of the democratic ideal and religious ideal, the classical capitalist ideal has gained a hallowed aura, beyond its intellectual and practical content. This is why the self-regulating market, the free market, has gained the greater share of theoretical development and attention than the industrial factory portion of the system.

Without the industrial factory system and the continuing improvement and innovation of new technology, the industrial capitalist system could not exist. However, within Britain and the United States, the pursuing and accumulating of money-wealth has always been a passion, surrounded with powerful social approval. Hence, the continuing emphasis on entrepreneurial practices and market dynamics in the puritan–protestant, legal-democratic Anglo-Saxon world.

The Industrial Factory System and Labour Antagonism in the British–American Model

Relations between factory-owners and the workers needed to 'tend the machines' emerged within a specific religious and economic context. As Calvinism pervaded England – the Anglican church accommodating much of the Calvinist ideology – and engulfed the USA, the puritan notion emerged that the capitalist factory-owner, in extending money wages to his workers, was setting them on a course for heavenly grace. For, he was paying them money, not enslaving them. And, in this same vein, the expansion of the use of machines to do the heavy work was also seen as religiously good, for if machines could do the heavy work, slavery could be eliminated. Thus, working hard for money while tending machines was seen as religiously purifying.

That is, it was seen as religiously good by the factory-owners. The workers, of course, felt differently, for the wages were not high enough to live on – in this world – and the work was too hard, too long, and often dangerous.

So, the factory-owners believed they were conferring heavenly grace, while the workers believed they were living in hell.

In terms of economic theory, labourers were considered as interchangeable parts – like cogs in a machine. Workers should move wherever the work is most lucrative, and thus free movement of labour would positively effect wage levels.

Smith and then Bentham wrote of utopian competitive economic systems, wherein factory production would create a huge economic benefit for all – the business-owners and the labourers alike.

Unfortunately, Dickens's England emerged in place of Bentham's. The condition of the workers became deplorable. Both living conditions and working conditions deteriorated dramatically from their serf–rural pre-condition. The economy did succeed. The industrial revolution was wonderful. The British did create a whole new world of factory production and urban living. The long-run results would be remarkably good. But what a hideous world emerged in the short run.

The reader is familiar with Dickens's description of the hellish poverty, the unhealthy working conditions which left workers maimed and spawned twisted children – twisted both physically and mentally.[6] The crime, vice, violence, and disease that emanated from the workers' slums, and, the lack of upward mobility, were hardly predicted by Smith or Bentham. Nor would Calvin or Tyndale[7] have seen any road to heavenly grace in these conditions.

From the world described by Dickens arose the history of labour antagonism to capitalism. Socialism, communism and unionism would arise to help the workers improve their lot, while the industrial capitalists in Britain and the USA would remain self-righteous within their Calvinist and Smithian world. Since the industrial capitalists believed their way was the sanctified way, they were unsympathetic to the plight of the labourers – they tended to blame them for their sloth, drunkenness, licentiousness, lack of drive and lack of education.

The socialist intellectuals then came to the rescue. They helped to organize the workers with unions and electoral parties. They demanded higher wages, better working conditions, and the right to vote – with no property qualifications.

As is so well known, from the 1850s to the 1930s, in Britain, the USA and Western Europe, worker–industrialist, socialist–capitalist antagonism became endemic in both the economy and the polity.

In Britain, this conflict was never resolved. Even today, the unions and the Labour Party are hotly antagonistic to the industrial-capitalists. Union obstructionism is legendary in Britain, as is the capitalist's stigmatization of labour as somehow less than human, or lacking in 'breeding'.

In the USA the attitudes were the same during the nineteenth century, but the explosive quality of the American economy allowed for rapid wage-gains for the workers. And, perhaps more importantly, the rags-to-riches, free public education, egalitarian-democratic ethos of the USA allowed for rapid worker upward mobility into the middle class. Thus, though the awfulness was there, its effects were ameliorated.

Today, in the 1990s, as the American economy 'downsizes' and globalizes, labour–corporate antagonism is rising again – and a good deal of violence is emanating from this conflict. Murder in the workplace is beginning to overtake industrial accidents as the number one cause of death on the job (1996).

We shall contrast the conflictual relationship between labour and capital in the Anglo-Saxon workplace with that of labour cooperation in Japan and Germany, shortly.

Labour conflict, worker poverty, socialist ideology or not, the industrial capitalist system and theory developed dynamically in Britain and the USA. In fact it was unchallenged, until the First World War and the loss of empire ruined Great Britain, and the depression of 1929 engendered economic collapse in the USA.

The Great Depression and the Keynesian Revision

With the great depression of 1929, the theory of the self-regulating market came into question. Would a socialist system take its place, was government control the only solution, would liberal democracy be lost with the statist solution?[8]

Keynes to the rescue: by creating a specified, rather than all-encompassing, role for the government in terms of intervention within the capitalist system, Keynes 'saved' the system and the theory.[9]

The government would invest in industry, infrastructure, and the development of technology. It would invest in public works supportive of both industry and the civic good. The government would guarantee production and create jobs. It would back away when the economy was strong, and intervene when the business cycle turned down.

The government would also regulate banks and insure their deposits in order to create confidence and stability in the economy. The stock market would be carefully regulated to avoid the level of total collapse that had occurred in 1929. Thus, the system would remain essentially capitalist, with the government playing a supportive role, rather than an obstructive or dominating one.

Keynes saved the system, but he violated the theory. There was too much government involvement from the classical–capitalist perspective. Given the alternative, however, of socialism and communism, a general, though grudging acceptance of Keynesian capitalism emerged in Britain and the United States from the 1930s to the 1970s.

The Cold War and Military Keynesianism (1945–1985)

The Second World War created the level of government involvement in the economy that Keynes had described. It was the war effort that really legitimized government support, more than the efforts that occurred during the depression era. This legitimization, in Britain, but especially in America, might have quickly been withdrawn after the war ended but for the fact of the cold war with the Soviet Union.

In the USA the cold war demanded an ever-greater military buildup, with ever-more sophisticated technology needed for military weapons systems. The race was on between the Soviets and Americans for nuclear weapons, missiles, space satellites, and so on. It was this unprecedented and highly technologized military-industrial expansion that inadvertently legitimized the Keynesian programmes.

Conservatives, who would have opposed Keynesianism, instead demanded an even greater military preparedness. Classical economists, who would have insisted on a return to *laissez faire*, kept quiet on policy and wrote only for academic journals.

It was military Keynesianism, then, that guided the American economy toward dramatic expansion, technological breakthroughs of world-altering proportions, and relative stability. Yet, military Keynesianism was never really accepted for what it was and what it accomplished. Instead, the massive government role was almost ignored. American businessmen and economists spoke as if the economy were running on purely classical principles, and as the cold war wound down, they began to insist on a return to a *laissez faire*, which they believed was close in principle to what had occurred in the 1950s and 1960s anyway! The role of the government in creating and disseminating technology, capital and guidance was never acknowledged.

The Death of Communism and the Revival of Classical Capitalism

With communism dead and socialism modified into legal-democratic welfare state capitalism, and, with the memory of the great depression fading, Keynesianism has declined in the USA and Britain. A revival of more classically-oriented theory and practice has emerged. The Thatcher policies in England and the Reagan revolution in America engendered a flurry of de-control, government withdrawal and entrepreneurial activity. This was accompanied by a theoretical reorientation.

Theories such as monetarism, neo-classicism, neo-Keynesianism, and others have emerged.[10] However none of these theories has been fully established, nor have the policies generated by them been fully successful. In fact, the economic realities of the new high-technology industrial capitalist economy seem to have outrun both the Keynesian and the classical vision.

Nonetheless, the more classically-oriented theories remain popular in

Britain and the USA, because they support the tradition of the non-interventionist state and the self-regulating market. As we shall see, the Europeans and the Japanese continue to view economic theory and industrial practice differently and, in all probability, given the heated competition from Japan, the EU, and now Asian capitalism, the governments of England and the USA may be forced back into a more Keynesian role – especially in terms of subsidizing technological development, protecting high-tech industries, maintaining an industrial base, guaranteeing fair global trading, and preventing stock-market fraud.

Even with a more Keynesian activism, the American and British model will remain more capitalist and less statist, more capitalist and less socialist, more capitalist and less bureaucratic, then its EU and Japanese counterparts. Further, the emphasis in this model still leads toward the enshrinement of business and finance over industrial rationalization.

In terms of politics – and this book is, after all, about politics – the more purely capitalist orientation of Britain and the United States fosters legal democracy. It is crucial to make this central: the British–American tradition of the minimalist state and the self-regulating market undergirds democracy by creating a separation of powers between the economic and political spheres. As Keynes put it, 'it is better for the ambitious man to tyrannize over his bank account then over his fellow citizens'.[11]

On the other hand, the continuing and widening wealth inequality generated by the Anglo-Saxon model allows the rich too much wealth and influence, drawing the polity toward plutocratic oligarchy, and gives the poor too little money or support, driving them toward crime and violence. Over 20 000 murders occurred in the USA in 1995! While in 1985, wealthy financiers bilked the stock market and the banks of billions of unearned dollars.

Only the size and prosperity of the middle class in the USA stabilized this volatile situation. And now, footing the bill for the indiscretions of the rich and the poor, and feeling the impact of global competition, even this venerable American middle class is beginning to weaken – shored up only by the entrance of the middle-class wives into the work-world. But, though the 'dual-career couple' saved the middle class financially, it may have destroyed it socially: the divorce rate is upwards of 40 per cent, while drug-use, family violence and family dysfunctions are on the rise.

Part of Britain's middle class – also stable and stabilizing – has emigrated to Canada, Australia and the United States, generating a 'brain drain' and a social malaise, which the arrival of Asian, Caribbean and African immigrants has only complicated.

Strengths and Weaknesses of the Anglo-American Model

The strength of this model is its very creative entrepreneurial spirit, which drives the system toward business innovation and the active application of every technological advance to the industrial process. In fact, it must be emphasized that the high-technology industrial revolution itself was created by, and fostered by, the American system. This creative strength, therefore, should not be underestimated.

Politically, as mentioned, the separation of spheres allows for the democratic state to operate with less control over its citizens and, with less of a bureaucratic apparatus, the democratic state relies more fully on the order and rules created by the law. Constitutional law and criminal law are both strengthened and legitimated, where the administrative bureaucracies of the state are reduced in reach and authority. The American fear of big government, is in this context, a very positive factor, in that it denigrates bureaucracy and elevates law. England sanctified law and legal authority, and America enshrined it in a living constitution. Neither the European nor the Japanese models utilize or legitimize the law in the same way as the Anglo-American model. Law and democracy reinforce one another, as Aristotle pointed out: 'A government by law is a government where reason and god rule, a government by men adds the character of the beast.'[12]

The weaknesses of the British–American systems, however, are also profound. Free-market economics, left to itself, swings through wild periods of boom and bust, engenders systemic underemployment, cares little for its workers, and engenders too great a disparity between rich and poor. All of these economic weaknesses produce terrible social consequences.

Inflation also plagues the system, and neither the classical nor the Keynesian theories explains it adequately.

Thus, even though a large middle class has emerged with the Anglo-Saxon model, and this middle class has served to stabilize the legal-democratic system, inequality, inflation, unemployment and violent crime continue to mar the Anglo-American dream.

GERMAN CATCH-UP INDUSTRIALIZATION: STATE PLANNING AND WELFARE PROGRAMMES

Britain's success in industrial production and subsequent rise to world power stunned the Germans. They had been the dominant military power of Europe and now were being by-passed. Weber's question, 'why the industrial revolution in England?' reflected the general German surprise and jealousy. Wanting to maintain their political power, the various German states, led by Prussia, embarked on a programme of catch-up industrialization.

Unlike the British industrialization, the German occurred with the kingly-bureaucratic state not only intact, but in charge. Bismarck, with the blessing of the Kaiser, organized his military-style bureaucracy with orders to aid a rapid German industrialization. Thus, the state was involved from the outset, not in monarchical meddling, obstruction and luxuriating, but in plan-rational industrial expansion.

Why the success, as opposed to the Stalinist-style failure? Because Germany, unlike Russia, possessed a strong and still vital tradition of trade capitalism, including institutions, ideas and active individuals.

The Hansa cities of West Germany,[13] the Low Countries and Scandinavia had been highly successful during the guild–capitalist era. Merchant families and banking families such as the Fuggers had been as legendary as the Medicis. Banks, insurance companies, businesses of all kinds, trading networks – all these existed in Germany, and the protestant reformation in its Calvinist form had legitimatized money-wealth, while in its more general Lutheran form it had at least made 'work' central to its ethic.

Therefore, when the Prussian state attempted to establish and expand industrial factory production, talented, energetic businessmen and venerable capitalist institutions were available to facilitate this industrial build-up. The businessmen and bankers were delighted to work with the state in this capitalist, yet patriotic, programme.

Real government–business cooperation occurred. The state created and helped found giant industrial monopolies in heavy industries, such as steel and chemicals. However, these industries were staffed with prominent businessmen, who operated them according to market principles. Efficiency, profit, pricing, wage levels – all of these were linked with market dynamics.

The government did not interfere with this capitalist set of processes, but rather attempted to aid the market dynamics through guaranteed bank loans for industrial expansion, investment in technological and scientific development – including university as well as industrial institutes – and finally, through tariff and other protectionist schemes, anathema to free-trading principles, but necessary for indigenous development in a catch-up situation.[14]

The combination of government–corporate cooperation worked in Germany. It worked because the government allowed and encouraged the business class to use market principles to guide the economy, and it worked because the government subsidized and encouraged technological science and rationalized industrial processes.[15]

The Russians did not do any of these things. Lacking an indigenous business class and trade capitalist institutions, and, believing them to be unnecessary and evil (anti-socialist), they attempted to eliminate the market mechanisms and administer the economy bureaucratically. As Weber predicted in 1918,[16] the Soviet bureaucratic economy would be inefficient, trapped in a maze of red tape, and dictatorial. Even when the Russians actively encouraged technology and science, they only utilized it for military production, neglecting civilian production almost entirely.

What about German labour? A socialist revolution was a real possibility in Germany. Remembering the violent peasant revolts of the sixteenth century, and feeling the fervour of revolution exploding again, Bismarck attempted a strategy of cooptation of the working class. He had the government extend an elaborate programme of welfare state measures to the workers, along with improved wages. Pensions, unemployment insurance, health insurance, safety measures, and good wages – all these helped gain the cooperation of the workers and their unions in the rapid industrial build-up.

And, if socialism continued to have hot appeal in Germany, it was the electoral sort – the Social Democrats – that gained ground, while the revolutionary communists began to lose their appeal. Thus, Socialism or not, labour–business cooperation became a hallmark of German industrialization.[17]

Industrial Rationalization and Patriotic Production

Two more elements unique to German industrial capitalism need to be described – not only because they differentiate German capitalism from British and American capitalism, but because the Japanese copied them.

First, there is the German craft tradition. Craft production was well developed in Hanseatic Germany. This craft tradition was applied to the industrial process and products of Germany. That is, craft pride and craft skill were directed into technology. This had not occurred in Britain and the USA, where the mass production of cheap products, to gain the highest profit, was the goal, and where craft production was obliterated.

In Germany, because German products came later to the market, an improved, better quality product, might help gain market share. Therefore, with craft skill as an indigenous advantage in Germany (and in Holland and Scandinavia), the tradition was re-channelled into the modern economy in terms of pride in workmanship and product perfection – even if the cost was raised somewhat.

This is not pure capitalism, and it created a trade-off in terms of volume and profit. Even today, Mercedes-Benz is priced too high because of its craft-like engineering. Yet it sells well world-wide, because those who can afford the more perfect product will buy it.

The second unique factor is the 'patriotic' motivation of the capitalists, workers and government bureaucrats. Since German industrialization was directed at catching up with and surpassing the British, it took on a patriotic tinge. This continued through the world wars, and still exhibits itself in Germany. The German 'volk', the German 'state', the German 'nation', were all invoked to increase productivity by labour and capital, and this patriotic current still pervades Germany, though publicly subdued since the Nazi horrors.

German Economic Theory: Protectionism, State Support and Technological Development

The work of the German economist List is barely known outside Germany and Japan.[18] List wrote early – just after Smith – he even advised Alexander Hamilton to adopt protectionist measures against British industry in order for American industry to develop. List openly advocated tariffs on foreign competitive goods and state subsides for indigenous industries.

Though out of the mainstream of classical capitalist theory, List's works encouraged – and described – the actual priorities of the American, and then the German, industrialists. Not only would the Germans heed List's prescriptions, but so too would the Japanese.

The other non-classical element in German economic theory was described and emphasized by Schumpeter.[19] The latter – one of the great historians of economic development – emphasized the centrality of technological innovations, breakthroughs, and adaptations as undergirding the remarkable and dynamic growth of industrial capitalism.

Accepting the creativity and efficiency of entrepreneurial business and market processes, he added the fact of technological innovation – in the industrial process and its products. This had not been emphasized by the market economists. For them, supply and demand, pricing and industrial expansions – all emerged from market competition and capital supplies. Industrial productive technique and technology were taken for granted.

However, economists like Veblen[20] and Schumpeter realized that the technological innovations altered the system in a causally separate sequence, and that both capitalist and industrial productive sectors influenced economic growth and change.

This emphasis on technology, technique and engineering skill pervades German economic theory and practice. So too does it absorb the Japanese.

In the Anglo-Saxon countries technological innovation is actively developed and introduced, of course. In fact, the USA still leads the world in technological innovation, initiating the 'high-tech' revolution by inventing and rapidly deploying the computer, the micro-chip circuit, the transistor, and television–telephone–satellite and cable communications systems.

Yet, surprisingly, the emphasis on technology remains absent from the theoretical economic models developed in the United States and Britain, even now, while the whole economic world is being caught up in the high technology vortex. Remember, though, that the classical capitalist dynamics still apply within the high-technology economic nexus, and that this portion of the theory is still developed almost exclusively within the English-speaking world.

Finally, another key difference, in practice and in theory, is the active role of the state in both industrial planning and welfare state support.

German industrial planning is institutionalized. The government planning boards – in Germany, and now in Austria, Scandinavia, and other EU countries – have an excellent record and reputations. They are simply accepted as part of the modified capitalist tradition developed during the catch-up era. The government planning agencies still work cooperatively with industrialists, business leaders, and banks. They help direct and regulate the economy, they control banks, they subsidize R&D and worker-training programmes.[21]

The success of this model in Germany and Scandinavia and the EU in general, has provided a challenge to British and American theory and practice, as Lester Thurow has argued.[22]

Along with the industrial planning boards, an extensive – and exemplary – welfare state has been created in Germany, mirroring that developed in Scandinavia. The welfare state programmes, such as health care, child care and pensions, are well administered and relatively efficient as state programmes go, providing 'cradle to grave' security and safety nets for those affected by capitalist dislocations.

However, these welfare programmes do add power and authority to the state, and therefore are both worrisome politically, while humanistic socially. The EU and Scandinavian programmes are so good that they have been emulated successfully in Canada. The USA remains 'exceptional' – to use Lipset's[23] term – in not emulating these programmes. If, however, the world emulates them – as in Singapore, for instance – will these programmes, in adding to the authority of 'statism', undermine the authority of law and representative democracy? Neither law nor democracy are popular in Singapore, for instance, at present.

In this treatise we are not suggesting that one model of modern capitalism is better or worse than another. Only that the different models of economic organization impact political democracy differently. We are attempting, then, to analyse the economic differences, and their political consequences.

Strengths and Weaknesses of the German Model

In terms of politics, the increased role and power of the German state does pose a potential threat to legal-representative democracy. If planning boards gain power and prestige in directing the economy, they gain legitimation in directing other areas of social interactions.

At the moment, the parliamentary institutes of Germany, the Netherlands and Scandinavia are functioning quite well. The planning boards have not overridden the parliaments. However, European political analysts are nervous about this and do monitor this potential problem. The institutionalization of the 'ombudsman' system,[24] for instance, was an attempt to check some of the power of the German-style administrative state. It is not clear yet whether this ombudsman system is effective. We shall analyse it more fully later on.

On the positive side, in terms of democratic stability, the German model has a distinct strength. That is, through its elaborate system of 'transfer payments'[25] – or heavy taxation on the rich – the middle class and the working class are made more secure. For the transfer payments from the rich are earmarked for subsidizing the free educational, family, medical and worker-training programmes.

The middle class is strengthened by these programmes, the workers have been gaining lower-middle-class life-styles, and the rich have been limited in their wealth accumulations, to some extent.

The resulting increased economic equality has generated a middle-class-oriented society, with far less of a skewing of wealth than exists in the USA and Britain. Neither rich nor poor flaw the democratic processes, and this is a definite plus for democracy.

Now, in Germany, though the class differences have been minimized, the influx of millions of foreign 'guest workers' has inflamed the kind of ethnic conflicts that once tore Germany apart. It is this ethnic tension, rather than class tension, which Germany will have to cope with, if it is to remain stable.

Another political problem created by the institutionalization of the administrative state in the German model is that bureaucratic authority often overrides legal authority. The welfare state is administered, the economy is administered – much of the polity is administered, instead of being legislated. Germans and Scandinavians seem more prone to turn to an administrative protective institution, like the ombudsman, rather than to the law courts. If Americans are far too litigious, Germans are not litigious enough. The law and the lawmakers are not fully legitimated within the German system.

In terms of the economic system itself, the weakness of the German model is in its lesser emphasis on entrepreneurial creativity. The kind of innovative, flamboyant changes created by American businessmen are lacking within the German system. The Germans tend to be conservative in terms of banking, business and finance.

On the positive side, the Germans are world leaders in science, engineering and technology, and have been excellent at generating assembly-line processes, and better-crafted machine tools, and other products. Excellent training and upgrading programmes for workers have also emerged from the combination of craft tradition and socialist-union power.

Conclusions

The German model of state industrial planning, welfare state programmes, and worker training and cooperation has become the model for the entire EU, and will probably become the model for Hungary, the Czech Republic, and other Eastern European nations. If it does become the European model, then all the strengths and weaknesses in terms of its effects on democracy must be carefully monitored.

As we shall now show, this German model became the basis for the Japanese model, which was to evolve even further away from the British and American *laissez faire*–minimalist state system.

THE JAPANESE MODEL: THE 'GUIDED' MARKET

All societies are unique, but most fit into certain basic comparative patterns – civilizational constellations or structural stages that typify an historical epoch. Japan, however, is so unique that most scholars have considered it a peculiar case – a case that could be fully understood, but not linked into any mainstream trends.

Now, however, with Japanese economic productivity becoming legendary, and with other Asian nations copying the Japanese system to varying degrees, scholars have been forced to study Japan, which in its economic structure may be peculiar, and yet, trend-setting, at the same time.

For us, in this book, however, what is more significant than Japanese productive efficiency is the concomitant process of Japanese politics. For Japan's parliamentary democracy is dominated by its economic bureaucracy. Most political scientists rate Japan's level of legal-representative democracy as less effective than its European and American counterparts.

If this is the case, then the key question becomes: if the Japanese economic model becomes more generally adopted by the mainstream of industrialized nations, will this produce a general decline in the authority and prestige of the parliamentary-democratic system of politics?

The answer to this question is central to this treatise. Therefore, it is necessary to examine Japan's culture and structure in order to sort out what will and will not become a world trend.

The Pattern of Consciously Copying Cultural Improvements and 'Japanizing' Them

When confronted with Hyksos chariots, the Ancient Egyptians – after a terrible defeat – copied them and improved upon them. The shock of military defeat and the desire for independence forced this process upon the Egyptians. So too the peoples of the Aegean and Middle East traded in their bronze for weapons of iron. Adopting and adapting superior technology is certainly not unique to Japan. What became unique, however, was the long-term, highly conscious effort made by the Japanese to carefully copy external technologies, while at the same time tenaciously attempting to maintain the internal integrity of their cultural patterns.

China, one of the great and advanced civilizations of the ancient world, sat across the water from the more primitive Japan. Cultural achievements, such as writing, sophisticated pottery-making, silk clothing, mystical and ethical world religions, and much more, found their way to Japan. Most cultures borrow from their neighbours, and so what the Japanese did relative to China was not out of the ordinary. What was unique – and continues to be unique – was (is) the organized, conscious pattern that evolved, whereby cultural improvements emerging out of China were carefully studied and systematically adapted to Japanese conditions. The careful study and the 'Japanization' of the foreign cultural institutions are what make Japan unique.

The existence of China, as a major mine of cultural enrichment, undoubtedly helped establish this peculiar pattern in Japan, along with its island-style isolation.

But the colossus China, aside from engendering Japan's copycat culture, had another profound effect upon the Japanese mentality. For, not only was China a fount of civilizational achievements, but after its conquest by the war-oriented Mongols, China, with its vast population, became a great military threat as well – both to the world and to Japan.

Japan's Real and Ongoing Fear of Military Conquest

Not only did Genghis Khan sweep across central Asia and terrorize the steppe-lands of the Black Sea rim, not only did the Khan threaten Europe with military domination by his 'Golden Horde', but, unknown to the Western World, the Great Khans threatened Japan.

Japan was a very militarized society, with a sizeable population. But Japan was divided between warring feudal clans and factions, and, even if united, no match for the myriad minions of Kublai Khan.

The Japanese were, at first, unprepared for the vast invasionary force that the Great King of China organized and sent to conquer them. However, as every Japanese schoolchild learns, the Japanese were saved from Chinese domination by the Kamikaze – the wind-storms 'sent from heaven', which destroyed the Chinese fleet. Not once, but twice did the Kamikaze winds save Japan from imminent conquest.

Having twice been saved by 'heaven', the Japanese decided not to risk a third heavenly intervention. They began what would become another deeply imbedded cultural pattern: a conscious, coordinated effort to learn and implement the most advanced military technology, and to create a military preparedness that would make Japan impervious to conquest.

The pattern of studying foreign inventions and Japanizing them was continued, but at the same time the Japanese pursued an almost paranoid policy of closing off Japan from foreign control. Since China remained a great fear-inspiring neighbour, this pattern persisted.

The Japanese fear of the Chinese colossus would become superseded by the sudden appearance of the European powers, who, as with the Chinese, possessed superior technology over Japan, and whose motives seemed clearly aggressive.

The European Powers as Another Set of Colossi to be Studied, Copied and Feared

China, the greatest power in the world, was being overcome by a new power – would Japan be next? The Philippines had fallen, South East Asia was falling; surely Japan would be on the list. From the first appearance of the Portuguese in the 1500s, to the appearance of the Dutch, English and Americans, through to the 1900s, Japan's national fear was heightened, and her national consciousness turned toward the adaptation and Japan-ization of Western cultural and military advances.

Just as writing, pottery, political organization and religion were copied from China and Japanized, so too the military, economic, political and ideo-logical forms of the European countries would be consciously copied and systematically adapted.

The goal of the Japanese remained the same from ancient to modern times: keep Japan strong by introducing every foreign technological, ideo-logical and organizational process from foreign nations, and, keep Japan 'pure' by retaining as much of the original Japanese culture as possible.

The Japanese Political System: Samurai-Mandarins

The history of Japanese political development is well known, and it is not our intention to repeat it, but rather to apply it to the contemporary scene to show what is unique and what is not.

From China, feudal Japan borrowed the emperorship. And, although this act officially united Japan into one kingdom, the actual power of the nation remained in the hands of the warring feudal samurai lords. How-ever, centralization did eventually occur, with the power devolving into the hands of the most powerful lord, the shogun, while the other lords were held hostage at the emperor's court for part of each year.

The emperor, though remaining a figurehead, did serve to unite the nation by becoming a 'divine' symbol of the nation and the Japanese 'race'. The 'yamato race' and the nation of the 'rising sun' became identification terms representing bonds and unique qualities of the Japanese people, in the same way that the German word 'volk' evoked these in the German people.

Notice that the Japanese copied the kingship from China, but though its form was the same, its functioning was completely different. The same would occur with parliamentary democracy, copied from the English and Germans later on.

Along with the divine kingship, the Japanese borrowed the Chinese system of national officials. The famed 'mandarin' bureaucrats of China were modelled by the Japanese, and, over time, the examination system and the Confucianist world orientation of the mandarins did emerge in Japan.

These government officials in Japan, however, were drawn in their origins and in their long-term historical recruitment, from among the samurai families that still retained great importance and power in Japan. With so many generations of samurai lords holding the administrative offices of state, it was natural for some of the samurai ideology and process to rub off on the offices of state.[26]

Thus the Chinese mandarin ethic, which evolved over the centuries in China, to emerge as the Confucian ethic – the ethic of the earnest literati-official – as Weber characterized it[27] – evolved in Japan to include a 'samurai ethic' merged with the Confucian-literati-oriented ethic.

So it was that the Japanese state bureaucrats evolved as part mandarin, part samurai. This hybrid survived and propagated itself, so that this patten has been maintained through to the modern era.

Again, this pattern is not unique to Japan, for, the German officials under the Kaiser – from Bismarck's time through to the Nazi era – retained the military character of the feudal (Junker) lords, from whom they had originally been recruited. There is, in Germany today, something still quite reminiscent of the military-style bureaucrats of Bismarck's day. However, while Germany's bureaucrats were famous for their efficiency, they were also infamous for their 'officiousness'. Weber called them the 'little dictators' who ruled 'without soul'.[28] This was not the case with the Japanese officials, who ruled with the grace of the Chinese literati, rather than the gruffness of the Prussian line commander.

From the Chinese mandarin tradition, the Japanese copied the system of the loyal, honest, intellectual and cultured official. The Chinese mandarin official was well-educated in writing and numbers and highly cultured in the arts and literature. They were a class of literati, rather than of dull bureaucrats. In fact, the Western term of derision for spineless, officious administrators – 'bureaucrat' – has no counterpart in the Chinese conception at all.

The exemplary Confucian official was the model the Japanese adopted. And, in Japan, as in China, the status of the state officials was very high. Since, over the years, the majority of the officials did hold the good of the community above their own self-interest, and did serve with integrity, the high status of these offices created an almost legendary reputation for these Japanese mandarins – they came to be seen as keepers of the nation's traditions and defenders of the nation's strength.

The samurai portion of the Japanese official is equally as important as the mandarin portion for an understanding of modern Japanese politics and economics. The Japanese officials – in the samurai portion of their tradition – see themselves as responsible for the strength of their nation and for its defence. They are constantly and painfully aware of the strengths and, or, advantages of foreign nations. They envision Japan as in a condition of potential war, wherein they must not fail to appropriate and adapt every advantage created, that could overwhelm Japan, or leave Japan vulnerable. The Japanese samurai-officials consider it their duty to guide the Japanese nation toward growth and strength and technological superiority. First, through industrial competition (in the nineteenth century), then through military competition (in the early twentieth century), and now through industrial competition again, the Japanese sought and seek, to keep Japan strong and invulnerable. If expansion into world markets is necessary for strength, expand; if protection of the home market is necessary for growth, protect.

For the samurai-officials, there is always a 'war-mentality' – a life-and-death competition in which the survival of the Japanese nation and 'volk' is everything.[29]

Before leaving this analysis of the Japanese state bureaucracy, let us extend our comparative analysis for a moment – for in other societies the state officials are not venerated, but feared or denigrated.

In the Turkish empire, for instance, the Turkish officials were charged with maintaining control over conquered populations. Furthermore, they had to tax the vanquished people, in money, in kind and in boys (to serve in the vast Turkish army).[30]

Therefore, the Turkish officials – operating in a sea of hostility and potential rebellion – developed a style of officiating dependent upon the firm establishment of power and punishment. Turkish bureaucrats were not only known for their competence or loyalty to the Turkish king, but for their cruelty and domineering style of administration.

This is not meant as any slur at all, but rather as an illustration of a contrasting style of administration. For, during the 1980s, when a team of Turkish sociologists questioned contemporary Turkish officials, they still found that they ruled through fear and intimidation, and that their establishment of power and domination over the population was their main goal.[31]

In the American system, which had no formal civil service system until the twentieth century, and wherein the best and the brightest talent is recruited into the corporations and the professions, government bureaucrats have come to be characterized as inefficient and dull-minded. If, in Turkey, the bureaucrats generate fear, in the USA, they generate derision.

In China and Japan the state bureaucrats generate veneration. And, this is very true today in Japan, though in China, the communist party bureaucracy may have disgraced and confused the tradition so much that little of the mandarin appeal is left. This remains to be seen. And, if the Japanese, Korean, and Taiwanese model of industrial production is fully adopted, the Japanese-style government bureaucracy may bring back the Chinese mandarin spirit. This latter may strengthen China economically, but as in Japan, it may inhibit the functioning of democracy.

Japan Copies German Industrialization in Process and Theory: Developmental Capitalism

As we have mentioned, the Japanese is one of the few cultures in history which consciously and systematically studied, incorporated and adapted outside ideas and institutions. Confronted by the European powers, as they had been confronted by China, the Japanese set out to copy their cultural advances. Unlike China, however, the European nations were not culturally homogeneous. Therefore, delegations of Japanese had to learn the different customs and languages of each of the European nations they wished to emulate.

At first fascinated with Dutch, English and French institutions – using the French army organization as its modern model – the Japanese, during their major and mighty effort to modernize, began to lean more and more toward the German model of industrialization and the German model of political organization. Let us analyse why this should have been so.

Germany, as we have already described, was confronted by an already highly developed English industrial system. Both the French and Americans were ahead of Germany as well. Therefore, the Germans, in attempting to catch up, created a system of government-supported capitalist industrial development. They not only implemented this system effectively, but they also created the theory to back up their practice. German economic theory – from List to Schumpeter – was, and is, different from English–American economic theory.

The Japanese made a conscious decision to copy the German model of rapid, catch-up industrialization. And, of course, certain similarities in the Japanese political structure and national ideology made this decision both politically easier and culturally more consistent, then if they had chosen to adopt the English or American systems of economy and polity.

First, Germany, like Japan, had a state bureaucracy attached to a monarchy. Secondly, this state bureaucracy had, in its top positions, individuals from a feudal military background. Thus, both the mentality and the pedigree of the samurai-mandarin was more like that of the German Junkers who were the officials for the Kaiser, than American or English civil servants. (Even where the English civil servants were of gentry origin, the gentry had lost their military-feudal orientation 200 years earlier, if we are to follow Tawney[32] in this.)

Thirdly, the Germans had developed the idea of the special nature of the German people and the unique character of the German nation: the spirit of the 'Volk'. Racial superiority, tribal unity and national destiny – all expressed in the German conception of the 'volk', mirrored exactly the Japanese conception of themselves: the superiority of the 'yamato race', the communal unity of the Japanese people, and the special destiny of the heavenly nation of the 'rising sun'.[33]

All these factors, which in retrospect of the Nazi horrors make us shrink away from any possibility of emulation, attracted the Japanese, who saw no moral atrocities at that time, but rather some spectacular economic achievements.

Along with this remarkable consonance between the Japanese and German politico-military structure and ideology, the Germans also provided the Japanese with a sophisticated blueprint for rapid industrial development. The German blueprint included these critical elements: a pro-active, military-oriented government bureaucracy, with high prestige and a goal of developing the national strength through industrialism; capitalist industrial firms organized as giant monopolies or oligopolies; investment banks owned, operated, or directed by the giant monopolies or the state; a labour force highly unionized and politically organized (by the socialists), but working cooperatively with big business and sharing the national goal of building industrial strength; an emphasis on industrial development, rather than financial profit.

The German economic theory that supported this process of industrialization provided the Japanese with another clear-cut set of policies for the inculcation of rapid industrial development, for a nation playing against stronger, better developed opponents. That is, the German theory advised protective tariffs for indigenous industries; direct investment from the banks, guaranteed by the state and guided by the state, into targeted industries designated for expansion; direct investment by the banks, guided by the state, into research and development of the most advanced technology possible – in terms of industrial production processes and the products produced; and government-directed and supported building of the national infrastructure – including everything from railroads to schools.

Now, it must be understood, as we have already made central to our discussion of Germany, that this German blueprint was, and is, a market system at its core. Unlike the communist systems that followed the Leninist–Stalinist–Maoist model, which ignored and ridiculed market economics, the German system followed market dynamics faithfully, but modified the market to more closely fit the imperfect conditions of an early developmental situation in a world where some nations were already highly industrialized. Even though the Germans and Japanese have continued to modify the market, they modify it by utilizing its own rules.

The Japanese Developmental State

From around 1868 onward, Japan became 'plan-rational' and developmental. After about a decade and a half of experimentation with direct state operation of the economic enterprises, Japan discovered the most obvious pitfalls of plan rationality: corruption, bureaucratization and ineffective monopolies.[34]

Thus, Meiji Japan – with no communist ideological baggage to predispose it against private property, and, with the private property models of Germany and England to study – began to shift away from state entrepreneurship to collaboration with privately owned enterprises. The government consciously favoured those enterprises that were capable of rapidly adopting new technologies and that were committed to the national goals of economic development and military strength. The collaboration took the form of close governmental ties to the privately owned industrial empires.

Having no big business firms in Japan of the scale necessary to create heavy industry, the government bureaucracy created large firms and staffed them with samurai family members whom they could trust would work for the national good. At first, of course, these large firms were operated directly by the government. But as the managers of the firms gained expertise in business and engineering, they gained more independence from the government.

The structure of the giant Japanese firms emerged somewhat differently from their German counterparts. The Germans focused their investments on heavy industrial monopolies and oligopolies. For, though this system flies in the face of Adam Smith's theory of perfect competition, it seemed the only way to catch up with English industry. Industrial firms of 'large scale' could raise and utilize millions in capital for technological development and modern production techniques. Small firms could not do this. This is the Veblen–Schumpeter portion of economic development, as opposed to the Adam Smith portion.

The Japanese copied this Veblenian–Schumpeterian model, and then went one step beyond it. Having entrusted the giant monopoly or oligopoly enterprises to a number of loyal samurai families, and wishing to avoid the conflict they feared could develop from business competition, the government bureaucrats simply commissioned the successful family monopolies to branch out and establish new industries in government-targeted areas of production. The new industrial units would be managed and improved within their own economic and technological necessities, but would be overseen by one of the samurai families that directed the successful earlier industries.

In this way, the Japanese created a system dominated by giant conglomerates which came to be called 'zaibatsu'. Today, these giant conglomerate units are called 'kieretsu', and their Korean copies are called 'cheobol'.

These zaibatsu giants are different from the German monopolistic giants in that they control many different unrelated industries. They are also different from American conglomerates – formed through financial takeovers – in that the Japanese units include separate management teams, expert in each separate productive area, though overseen at the top by the unified zaibatsu executiveship. American conglomerates have, by contrast, run into difficulties because they have consolidated management teams, and often fired specific expert managers in one or another productive area, in order to cut costs and pay back the exorbitant loans needed for financial takeovers. The Japanese giants do not operate like American conglomerates in this latter way, and in fact, receive more capital from their zaibatsu control centre, rather than less, whenever they need it for long-term development or branching out to new fields of production.

The zaibatsu, or kieretsu system, then, was modelled on the German giant monopoly model, and then developed in an even more concentrated direction.

It must also be mentioned that the Japanese government bureaucracy did attempt, in the early years of industrialization, to control the economy from the state down in a direct military fashion – as the communist regimes did later. However, the Japanese officials learned rather quickly that the zaibatsu business leaders and industrial managers had gained a kind of hands-on expertise that they – the bureaucrats – lacked. Since the government officials respected the zaibatsu executives and were personally connected to them in terms of the 'age-group' educational ties they had shared and their elite class background, developing an almost 'fictive kinship' tie, they tended to listen to them and even defer to them, in terms of the decision-making process.[35]

This is highly significant. For the Japanese economy would have developed as purely statist, and militarist. The Japanese might have developed the first 'command' economy which might have evolved like that of the communist systems of Russia and China.

But the Japanese economy did not evolve in this direction, but rather, away from it. The reasons for this are: one, the Japanese copied the German system, wherein the great firms held real independence from the government, though cooperating with the government in a very direct way and, two, the Japanese officials actually tried to direct the zaibatsu more completely, but noticed – to their credit, and as a testament to the unique Japanese cultural tradition of copying a cultural achievement correctly and improving upon it – that when they overrode the in-put of the industrialists, the economy became less efficient.

So impressed did the Japanese government officials become with the hands-on expertise of the businessmen, that they extended the process of 'on-the-job input' all the way to the shop floor! Not only did government bureaucrats listen to business leaders and industrialists, but they also listened to the assembly-line workers and engineers involved directly in the production process.

Japan, then, did not evolve toward a mindless, command economy – Russian style – but evolved instead, toward a system blending the best of entrepreneurialism with industrial planning. Input into industrial decisions was sought from every level of individuals active in the productive process.

Another capitalist economic institution, copied in its German form, is the banking system. Japan had no banks. It could have instituted a system like that of England or America. The German system was chosen because it was directly linked with industrial development, rather than financial wheeling and dealing. German banks worked with big industrialists to engender long-term growth. This is what the Japanese needed. Therefore, Japanese banks were created by the government as subsidiaries to the zaibatsu. These banks worked out a lending programme that would facilitate long-term growth and minimize interest repayment.

Similarly, lacking a stock market in Japan, the Japanese government created one. But again, the 'capital market' was tailored to long-term growth and direct support of zaibatsu expansion. Thus, the Japanese stock market does not function in terms of stockholder profits, or for speculation, but rather as an adjunct to the banking and zaibatsu development programme. Still, there is a stock market in Japan, and it can be extended in its financial functioning. Take note that there was no stock market in the communist economies.

Labour Cooperation in Japan

What is distinct about Japanese labour is its fully cooperative relationship with its zaibatsu industrial overseers. Rather than evolving toward a system of class antagonism, compromises were made by both business and labour which served to keep class conflict to a minimum.

There was, and is, a socialist party; there were, and are, unions; there were radicalism, violence and strikes; but mostly there was a spirited cooperation, more like that of elite army units or sports teams, and unlike the labour–management relations of the Europeans.

The reason for the cooperation and high morale of the workers lay in the system of 'clan-communal' caring and life-time employment which the Japanese established.[36] And, again, this did not emerge from Japanese cultural traditions alone – though these heavily influenced the process – but also from the 'corporatist' model of production developed by the Germans and Austrians.

Small Business in Japan

The small business system in Japan is also different from its European and American counterparts in some ways. It should be emphasized that small business does thrive in Japan. A huge network of retailers and middlemen exists in Japan, similar to that of the USA. What is different about the Japanese small business system is the network of very personal ties that exist between the small businessman and the zaibatsu.[37]

This would not be a significant phenomenon, except that foreign firms find that they cannot do business in Japan, because they lack the personal ties to the distribution network controlled by small businesses.

Since the Japanese government and big business do not want foreign competition to enter Japan, they do little or nothing to help foreigners understand their small business network, and they do less to encourage the small businessmen themselves to interact with foreign firms.

Industrial Production Emphasized over Business and Finance

The Japanese are obsessed with the improvement of the industrial-productive processes, and in love with technological innovation. Financial wheeling and dealing occurs in Japan, but it is less characteristic, and it rarely involves the core zaibatsu (kieretsu) industries.

Scientific management, or Taylorism, has become deeply ingrained in Japanese economic thinking. Japanese 'scientific management' has been very successful, in part, at least, because the workers are treated as dignified human beings with an expertise to share with management.

Japanese managers ask for shop-floor input from the assembly line workers. This makes the workers feel proud and smart, rather than alienated. Most importantly, since the Japanese guarantee life tenure to the workers of the giant firms, the workers encourage and welcome automation, and improve upon its shop-floor functioning. American unions and workers oppose automation, because they know it will cost them their jobs. Of course, American industry has automated, but this process has been accompanied by assembly-line sabotage, worker antagonism, unemployment and violence. In Japan, automation has been celebrated by management and unions alike, and implemented with great rapidity and success.

A Japanese scholar, Sahasi, likes to quote Schumpeter, to the effect that 'the competition that really counts in capitalist systems is not measured by profit margins, but by the development of new commodities, new technologies, new sources of supply, and new types of organization'.[38]

The Japanese, following Schumpeter, as well as Veblen and Taylor and other industrial 'efficiency' theorists, focused all their efforts on what they call, 'industrial rationalization'. By this they mean the rationalization of the industrial enterprises themselves, in terms of new techniques of production – from Ford's assembly line to high-tech robotization, investment in new equipment and facilities as an ongoing programme keeping pace with invented improvements, quality control – including technological improvements and labour cooperation, cost reduction, through more efficient productive techniques, better technology, and more efficient supply systems, and new management techniques as well.

The theory of industrial rationalization goes back to Veblen, in his *Engineers and the Price System*, and then, in more simplified and direct form, to Frederick W. Taylor's system of 'scientific management'.

As with Veblen, Japanese economic planning officials believe that market forces alone will never produce the desired improvements and expansion in industry. And, along with its commitment to free enterprise, private ownership of property, and market dynamics, MITI (the government planning agency) believes industrial rationalization and development are still absolutely necessary.

It is not for us to discuss in this treatise, nor do we deign to suggest that we have the expertise to do so, but given the Japanese emphasis upon, and success in 'industrial rationalization', it seems clear that this portion of economic theory should be pursued, along with the development of monetarism, fiscal policy, price theory, and other classical capitalist forces. After all, mathematical game theory and other mathematical paradigms cannot help improve shop-floor efficiency, or invent and apply new technology.

Capturing Market Share, Rather Than Maximizing Profits

You will remember that we have characterized the Japanese economic ministers as samurai-mandarins. This means that their outlook will, in part, be coloured by military overtones. Thus, Japanese economic strategy is geared toward 'capturing market share', rather than gaining maximum profit, and gaining the advantage in trade.[39]

Most remarkably, Adam Smith's basic principle – 'profit maximization' as the mechanism that drives the capitalist economic machine – is partially rejected by the Japanese. They are motivated by profit-seeking, but are willing to postpone profit-taking in order to capture a larger market share, and ensure long-term competitiveness.

Now, if we leave the realm of Smithian ideal behaviour, and turn to actual historical economic behaviour, we discover that American and European monopolies (trusts, cartels) used to take over markets by temporarily lowering prices and driving out smaller competitors. In fact, this is precisely the process followed by capitalist firms throughout American and European history. Of course, in the USA and Britain, the Smithian principle of business competition, and conversely, the Smithian warning about the inefficiency of monopolies, led to the anti-trust laws and anti-monopoly policies.

Nonetheless, giant oligopolies continued to dominate most major industrial areas through to the 1980s. And, only in certain new fields, such as electronics and computers, have smaller firms been able to survive. In Europe, the giant oligopolies function more closely to the Japanese model, than to the American. And, even in the USA the tendency toward concentration and oligopolization is ongoing (though the financial wheeling and dealing and hostile-takeover process generated by Reagan's de-control programme has obscured the process).

The Japanese, then, are not doing anything that has not been done in the other industrial economies, except that they openly state their policies and defend them publicly, and, they have held on to the catch-up mentality in their developmental programme longer than other nations.

The Japanese, through the 'dumping' of products and low pricing, flood the market and drive out their competitors – they did this in machine tools and electronics, and tried to do it with micro-chips. They also have been successful at driving out competitors in the auto industry. Remember Volkswagen in the American market? It is now only a memory. However, did not Standard Oil drive its competitors to ruin? And, what of Ford, GM, General Electric and other huge American firms? The difference, again, is that the American economic theorists and politicians cried foul, and the American government responded by trying to break up the monopoly economy. In Japan, the government does encourage internal competition among its giant firms, but encourages them to act with monopolistic intent in foreign markets. For the Japanese, this is an economic war, and they want to win it.

However, the Japanese need the American consumer market for sales, and are completely dependent on foreign sources of raw materials. Aware of their weaknesses, the Japanese know they are vulnerable in the economic war. This combination of strength and weakness is Japan's historical cynosure – it engenders the peculiar attitude of external aggressiveness with a defensive core.

The 'Guided' Market

It must be emphasized again, so that no connection with communist-style planned economics can be assumed, that the history of MITI is the history of its search for 'market-conforming methods of intervention'.[40]

The Japanese 'guided market', directed by MITI and the other economic ministries, in implementing their industrial policy, incorporate market principles at every level of economic activity.

Thus, the principles of supply and demand, profit, efficiency, trade balances, pricing and so on, are all meticulously studied by the Japanese administrators, and the market variables set the framework within which Japanese industrial policy is established.

The Structure of the Japanese State Bureaucracy

In Japan, the economic bureaucrats are central to the functioning of the system. The government agencies responsible for economic policy are the now famous or infamous MITI, or the Ministry of Finance, International Trade and Industry; the Ministry of Agriculture and Forestry, Construction and Transportation, plus an economic planning agency.

These government agencies attract the most talented graduates of the best universities in the country. Tokyo University (Todai) is the most prestigious, but there are other universities from which the officials are chosen.

The state ministries in Japan make most major decisions, draft virtually all legislation, control the national budget, and are the source of all major policy innovations. MITI is charged with the formulation of most industrial policy, and oversees the carrying out of such policies.

Policy change in Japan occurs through internal bureaucratic disputes, factional infighting, and conflict among the ministries, rather than through parliamentary debate.[41] This is highly unusual, as compared even to corporatist economic systems, such as Austria's, or socialist-oriented ones, such as Sweden.

For, even though the state bureaucracies of all modern nations are well developed, policy debates in other nations take place either within the parliament, within the political parties, or from pressure groups and interest groups external to the state. In Japan, these debates, conflicts, and policy decisions emerge from within the state bureaucracy itself.

The exact structure of this Japanese bureaucratic system is as follows: from grade school through high school, the young people are pressured to do well in school. Both boys and girls are pushed hard academically. During the high-school years, the boys and girls are pressured to take the examinations for the university. Girls are not pushed as hard – the Japanese system is still very male-dominated.

The examination scores determine which university the student will attend, the better scores gaining one admittance to one of the prestige universities. So far, the Japanese system does not seen different at all from the European system, which exhibits a similar set of examinations for entrance into the universities. However, once in the universities, the Japanese students go on a temporary 'vacation'. That is, after years of working hard in school and studying strenuously for the entrance examination to college, the Japanese students are given a temporary respite. They take courses, and do study, but they also are given time for romance and travel, and friendship-bonding. The college years are very relaxed.

Then, however, the pressure begins once again. After graduating from the relaxed college years, entrance into law school or graduate school begins. During the law school or graduate school years, the pressure is put on again, and increased 100-fold. For, after the four-year vacation in college, preparation for the big examination begins. This exam is called the Higher Level Public Officials Examination.

Competition is fierce, and the exam is extraordinarily difficult. Many – even most – do not pass it the first time.

Those who cannot pass the exam will be slotted for lower-level government work, or middle-level management in the big companies. Failure to pass the examination definitely limits one's life chances, and excludes one from the top leadership positions. Since, however, the examination is meritocratically instituted, those who fail blame themselves, not the system.

It should also be mentioned, that for those individuals who do not gain entrance to the prestige universities, the big exam is still available. Some few do pass it, and once in the government bureaucracy must then try to become part of the network of friendship groups that they had been excluded from. There have been success stories in this regard.

One wonders also about the suicide rate amongst those who do not pass the examination. Also, as yet, there has not been a major feminist push to enlarge female participation, though the number of women who take the exam and pass it has increased.

Once the exam is passed, entrance into the state bureaucracy begins.

The young graduates are then recruited into one ministry or another, based both on the needs of the particular ministry and the network of connections forged during the university years.

The Old Boy Network Begins at the Elite Universities

The Japanese call the cliques of classmates that form at the universities, 'gakubatsu'. These gakubatsu cliques are inseparable from bureaucratic life. For, it is the entrance into the elite universities, and the passing of the big examination that sets the bureaucrats apart from the common Japanese and links the young men into cliques.[42]

Japanese society, as is well known, is a society in which age gains one status. Seniors are held in high esteem and considered wiser and less impulsive than the young. Yet the energy and creativity of the young are also recognized in Japan.

Thus, the young, energetic, creative bureaucrats – linked in friendship clique networks of very intimate classmates and workmates – are also integrated into carefully institutionalized vertical networks of seniors.

'Age-grading', and respect for seniors, influence everything the bureaucrats do. And, remarkably – if one wishes to place this system in comparative-historical perspective – the age-graded, senior-oriented system of authority hearkens back to horticultural societies.[43] It is a post-tribal, pre-monarchical phenomenon, exhibited in other societies, and of course, indigenous to pre-feudal Japan.

What is so remarkable is that the Japanese have retained their traditional age-graded system, and adapted it to modern conditions with incredible effectiveness. So often, traditional modes of social organization act as 'fetters' – to use Marx's term – which inhibit progressive changes. Not in Japan! Here we have a traditional system, preserving the deference for elders, male domination, and age-grade clique groupings, which operates to create modern economic and political decisions in a most successful way. This combination of modern technology with traditional customs is very appealing to other Asian nations, such as Singapore and Thailand – will it have appeal in China?

Public Administration is Emphasized, Rather than Law, Business or Academic Economics

At the graduate schools – the most famous of which is Tokyo University Law School – the prospective bureaucrats study 'public administration'. This has its counterpart in France, under the French rubric 'administrative law'. But the Japanese officials do not study 'law' in the sense that this word is used in American law schools. For the scholar of 'legal-rational authority', this is problematical indeed. For, unlike most American government officials, the Japanese officials will not be steeped in the tradition of constitutional law and 'inalienable' individual rights, but rather, in a tradition of mandarin administrative and ethical authority, and the national good.

Not only are Japanese officials not trained in 'the law', but they are also not trained in the entrepreneurial business tradition either. Japanese officials do learn business principles, but they are not trained to focus on the profit motive, the individual accumulation of wealth, or the sanctity of business competition. They are trained in national industrial policy-making – including the indigenous development of industrial productivity and the patriotic necessity for success over foreign competition.

So, let us repeat what Tokyo Law School and the other prestige graduate schools actually teach the prospective officials: they offer a superb education in public administrative law of the continental European variety. They also study public finance, economic policy and industrial policy. They become very well grounded in what they call 'practical economic theory'. From our point of view – and given the Anglo-Saxon economic traditions, this is a heady oxymoron – the Japanese officials study market-based planning.

The Network Linkage Between the Government Bureaucracy, Big Business and Parliament: 'Descend from Heaven'

The Japanese system of early retirement from the bureaucracy, into big business, banking or parliament, called 'descend from heaven', is quite unique, and very functional.

First of all, since early 'retirement' takes place in one's fifties, this system allows for fresh new energetic minds to take over the functions of state on a regular basis. This system helps avoid bureaucracies' greatest problem – stultification and ossification. Weber warned over and over again, like a modern-day Jeremiah,[44] that the enlarged bureaucratic state could lead to a nightmarish ossification of functioning. To put it in modernized Kafka-esque terms, a nation could become like a giant motor vehicle bureau or post-office – functioning, but barely, and strangling to death on its own red tape and rigidity. The communist systems – as Weber predicted – were crushed under the dead weight of their own bureaucracies.

Therefore, Japan's innovation – retirement from the bureaucracy at around age 50 – is a truly remarkable improvement. For, no ossification can occur, with new age-sets of classmates entering government service, anxious to make their mark on the progressive development of the nation.

Yet stability is maintained in the system, because the older bureaucrats do not retire at all, but rather sit on the boards of the businesses and banks, and sit in parliament. And, though they are subject to the authority of the younger government bureaucrats, as seniors, they command great respect, and will be deferred to in many cases.

Thus, a very nice balance between adults and elders is maintained in the Japanese system.

Second, and perhaps more importantly for us in this treatise, as this factor relates directly to the functioning of the Japanese political system, the 'retirement' of the state officials into the big businesses, banks and parliament creates an 'old boy' network of leaders who dominate and direct the economic and political system. This Japanese network operates with far less of the separation and limitation of powers that typifies the western democratic systems – especially the American system. There is little of the structural limitation on power that Locke and Montesquieu believed essential, in Japan.

The Bureaucratic State Dominates Industry and Parliament

The Japanese system of retirement, into the big business, banking and parliamentary spheres, ensures that there will be excellent communication between the state bureaucracy, industry and parliament. This is good in terms of industrial planning, to be sure. The kin-like network ensures that business leaders will be heard – that their ideas and their needs will be fully understood by the officials. Parliamentary leaders will also gain a full hearing, making the needs of their constituents clear to the government bureaucracy.

But, it must be clearly understood that the bureaucratic state rules. The parliament has been called a 'puppet parliament', and, business leaders too, defer, in many cases, to the decisions of the state bureaucrats. This is not always the case, of course. The two most famous cases of business defying the bureaucrats are that of Sony and Honda. But, their success is the exception proving the rule. The same can be said for the recent elections which unseated the reigning LDP party, for the socialist prime minister ended up jettisoning their whole agenda in favour of that of the LDP.

More on these examples later. Here, let us emphasize that the Japanese system of state-bureaucratic domination seems to be, in general, good for industrial growth, but definitely bad for democracy.

A comparison with the United States process may prove helpful here.

In the United States the flow of power and influence is exactly the opposite of that in Japan. In the United States, the wealthy business stratum influences the process through a system of recruitment of its leaders from elite private high schools and colleges, law schools and graduate schools. The best and the brightest young men (and women, recently) are then recruited into the Wall Street banking and brokerage houses, Wall Street law firms, and Fortune 500 business firms.

Then the process runs in the exact opposite direction of Japan – that is, at retirement, the business executives, bankers, and top lawyers are brought into the government in cabinet positions, top bureau chief positions, and ambassadorial positions. Others are recruited to Washington, DC as lobbyists and campaign fund-raisers for both political parties. Still others are encouraged to run for the Senate – which has recently been called a 'millionaires' club', or in fewer cases, for the House of Representatives.

E. Digby Baltzel[45] has clearly spelled out this process of power and influence, flowing from the business upper stratum to the government, as have Domhoff[46] and Mills.[47] It is significant that Baltzel – a conservative favouring the American system, Domhoff, a radical, highly critical of it, and Mills – a Weberian, warning of the increase in bureaucratic power – all present essentially the same set of dynamics: all picture power originating from the business class and influencing the democratic processes of state. All, however, still consider the American system democratic, since electoral power can, and does, counterbalance business power. Even a pluralist theorist, such as Dahl, in *Who Governs*,[48] later altered his theory to include what he saw as the over-influence of *some* interest groups, i.e. big business, in an otherwise pluralist democratic system.[49]

In any case, few would disagree that in Japan, the flow of power is from the government hierarchy to the business and parliament, whereas in the American system the general flow of power is from the businesses to the federal government and Congress.

Now, the Marxists call the American process plutocratic oligarchy – or government by the business rich – 'bourgeois democracy' – and, of course, they criticize it, and rightly so. However, the Marxists miss a very essential point in their critique of 'bourgeois democracy'. The point they miss is that – in terms of the theory of the separation and limitation of power – the 'whig–liberal' theory of English and French Enlightenment vintage[50] – the American system, as it stands, weakens the power of the state in that the economic power is held by the business strata – financial, corporate, industrial – which are separated from the state.

Laissez faire, in demanding the limitation of the power of the state over economic enterprise, removes power from the state and vests it in the business class. Thus, if 'the state', in Weber's famous definition, is that institution which maintains a monopoly of power, then 'the state' in a bourgeois democracy is necessarily weakened by having the economic component of power removed from it, and banned to it.

With the monopoly of state power neutralized by the countervailing power of the business strata, electoral democracy becomes active in promoting the interests of the other groups in society, even though these may be less influential than those of the business strata.

The Marxists missed this point – they were too ready to denigrate Enlightenment political theory as bourgeois propaganda.[51] The Marxists paid dearly for ignoring the Enlightenment principles of the limitation and separation of powers. Even modern Marxists, such as Poulantzas,[52] made this error – and, of course, the whole Leninist-style political structure, in vesting all power in the state, ended in a hideous totalitarianism, rather than a utopic workers democracy (like the 'Paris Commune').[53]

But – and we are finally coming to the problem we wish to highlight – the Japanese miss this point as well.

Some Japanese scholars and journalists are becoming aware of the problem of the all-too-powerful bureaucratic state. They are using terms like the 'puppet diet'.

For, in the Japanese system, the state dominates big business and the parliament (diet). Thus, if the state became militarist and neo-fascist, which it did in the 1930s, the parliament would be unable to block the authoritarianism and brutality that emerge with it.

With no functioning system of 'checks and balances', and with no real separation of economics from political power, the Japanese system was vulnerable to fascism, and remains vulnerable to it.

Even today, the Japanese state officials act as if they were 'above the law'. Rather than the 'rule of law' prevailing, finds Henderson,[54] 'a rule of bureaucrats' prevails. If this is the case, and if most of Asia follows the Japanese model, will Weber's nightmare of a world run by 'soulless bureaucrats' become a reality? Will Mills' 'power elite' dominate the affluent 'subjects' of a new form of autocracy?

As Robert Wade[55] puts it, the state bureaucrats rule, the parliamentary politicians reign; the politicians don't make policy, their function is to create space for the bureaucracy to manoeuvre, and to act as a safety valve by forcing the bureaucracy to respond to the needs of groups upon which the stability of the system rests (such as small business, the farmers, labour, etc.).

This distinction between 'rule' and 'reign' engenders what Wade calls soft authoritarianism,[56] based primarily on the need to maintain economic developmentalism. This 'developmentalist' orientation still pervades Japan today as much as it did in the 1890s!

One-Party Domination of the Japanese Parliament

The emphasis on economic development, and the superordinate position of the bureaucracy in this process, has also pressured the Japanese parliamentary system into leaning toward one-party domination. For, though opposition parties exist and can be elected to power, the system runs more smoothly with the bureaucracy and one single party interlocked.

From the Meiji period onwards, and following the Bismarkian model of politics, though the parliament was controlled by the bureaucratic state, real political parties did emerge. Political parties, of course, did emerge in Germany, and became fully-fledged electoral parties: the Christian Democrats, Social Democrats, Monarchists, and later the communists and the fascist parties, all began to compete in Germany, even though the state still held an iron hand in policy-making.

This process was repeated in Japan, although the parties were weaker, in terms of real voter support and ideological fervour. In Japan, a capitalist-industrial party became the majority party. This party had close links to the bureaucracy and an electoral base among the small farmers and small businessmen. Again, as in Germany, a socialist party emerged and recruited support from the workers (linked to the unionization process, but with less militancy than their German counterparts). A communist party also emerged in Japan and would become important in the wake of the Chinese communist victory in 1948. In the 1930s, predictably, a fascist militarist party emerged, which developed deep roots in Japan and surprisingly strong commitment from its followers. This passionate military–fascist party took root – and still engenders loyalty – as linked with Japan's samurai–military tradition. If fascism – at its core – linked the Germans to their glorious Teutonic knightly past, and the Italians to their world-conquering Roman legions, the Japanese were reminded of their samurai tradition. This tradition runs so deep in Japan that the right-wing parties still are able to stir turmoil in the populace, in the 1990s, as they go around in their sound trucks blaring samurai-style propaganda. As is well known, many of Japan's most honoured intellectuals and novelists sympathize strongly with this proto-fascist ideology in modern Japan.

Thus, if there was less fervour on the left, there certainly was political disagreement and conflict in Japan. The political parties, in this sense, acted as real political vehicles representing different points of view, as Japan's 'civil society' slowly emerged from the 1890s to the modern era.

However, though the socialists and fascists existed, Japan was controlled throughout most of its modern history by the centrist industrially oriented party.

Japanese politics has not simply been one-party politics, however. One must be very careful in this analysis. For, during the late 1930s and through the Second World War, the fascist–militarists captured control of the government. Both the state bureaucracy and the industrial party were dominated by a group of political militarists and actual army and navy officers. There was a coup in the late 1930s, and right-wing assassins murdered some centrist leaders. Japan's version of fascism did emerge. The state and parliament did knuckle under to the militarists' demands for massive international imperialism and war.

Though there was nothing like a popular fascist movement – as in Italy, Germany and Spain – and, though there was no larger-than-life charismatic leader, such as Mussolini or Hitler – yet, in Japan, the Emperor himself was used, willingly, though this was later covered up by both the Japanese and the Americans[57] – as the national charismatic symbol. The Japanese fought and died for the emperor the way the Germans fought and died for Hitler. And General Tojo was the closest thing to a fascist leader which the Japanese would produce. (He actually was part of a large clique of militarists who collectively made war decisions in Japan.)

So, though it is true that the industrialist party and the state bureaucracy remained in power, from about 1936 to 1945, the semi-fascist clique actually controlled Japan. We are only characterizing Tojo and his clique as semi-fascist, because they did not totalitarianize or brutalize the Japanese population in the way that Hitler, Mussolini or Franco did. Be advised, however, that these Japanese semi-fascists did brutalize the South East Asians and Chinese, and the American and British prisoners of war taken in the Pacific campaign. They also forced the Japanese workers to work like slaves until they nearly died, from both overwork and under-feeding, from 1943 to 1945, when the war was being lost. And, near the end, they encouraged the mass suicide of Japanese battalion troops, through 'banzai' attacks and individual 'Kamikaze' acts of patriotism.

When World War II ended, the Americans attempted a sweeping democratization effort in Japan. Let us look at the American attempt and the results we helped produce.

The MacArthur Era: Anti-Fascism, Democratization and Anti-Communism

After the Allied victory in World War II, General Douglas MacArthur was empowered to create a sweeping democratization of Japan. His first mandate was to remove, try, and jail the military and semi-fascist leaders of Japan, who had been responsible for the excessive cruelty in China and South East Asia, and who had ordered the mistreatment of American and British prisoners of war. Prime Minister Tojo was executed, and many other militarist leaders were jailed.

However, the retention of the Emperorship – in its pristine form – also engendered a cultural conservatism, which inhibited parliamentary democracy and reinforced the traditional samurai–mandarin rule, which the Americans were trying to override.

After all, the Germany that was re-created after World War II had no more kingship. And the Kaiser's stronghold itself, Prussia, had been separated off by the Russians. This allowed West Germany to evolve in a much more democratic direction then it might have. West Germany became more like the Netherlands than like Imperial Prussia. There may have been a chance to move Japan in this direction, as well, but this chance was lost once the emperor was cleared and reinstated.

When China went communist and the Korean War erupted, the American government – still charged with the occupation, revitalization, and democratization of Japan – decided to take drastic steps to shore up the capitalist industrial party and build up the anti-communist forces within Japan.

The LDP – Liberal Democratic Party – was formed, and it was neither liberal nor very democratic. It has reigned since the 1950s in an unbroken electoral domination until 1994, when it finally lost an election.

This party is a capitalist–industrial party, such as the Christian Democrats in Germany or Italy, but it also retains elements of the semi-fascist military movement of the 1930s, and elements of the traditional samurai–mandarin leadership clique that traces itself back to the Meiji period of the nineteenth century.

Now it must be understood that the LDP has been good for Japan in terms of industrial productivity. No political party has a better record in this regard – not even the German Christian Democrats, who faltered during the late 1960s, when the LDP became more successful than ever.

In terms of democracy, however, the LDP's domination has not been so favourable a phenomenon. Elections in Japan looked more like plebiscitary affirmations than true contests involving differences of ideology or policy.

Japan's Political Parties in the 1990s

The socialist party finally won an electoral victory in the early 1990s. They won because of a number of financial scandals concerning the LDP.

However, having won, they were not able to govern without forming a
coalition with the LDP. Having formed the coalition, the socialist leader
who became prime minister then immediately repudiated the socialist programme and adopted that of the LDP! The people, he correctly assessed,
had voted for a cleanup of corruption, not a change in economic policy.

Having formed the coalition with the LDP, the new prime minister then
demanded a mind-boggling political change: he proclaimed that the socialist party should disband completely, and reformulate itself as a new kind
of opposition party.

This is hardly typical of parliamentary politics in most nations. But, in
Japan, it made sense. No party can rule there – at the moment – without
the LDP, because the LDP is linked with the state bureaucracy and the
zaibatsu (kieretsu). And the socialists have no relevance in a nation where
jobs are guaranteed for life, wages are excellent, bonuses are regularly
given, CEOs are hard-working and not overpaid (as are their American
counterparts), and so on. What would the socialists want that the workers
do not already have? Yes, workers in smaller industries do not gain the
benefits of the zaibatsu workers, and this should be corrected. But, this is
not enough of a problem to sustain a socialist party in Japan.

If the socialists do disband – and they may – will a new party emerge?
A party perhaps, representing the expanding middle class?[58] After all,
Japanese consumers do not lead as luxurious a life-style as their European
or American counterparts. Japanese pricing is set for export, therefore,
Japanese consumers are grossly overcharged so that the export products
can be sold for less abroad. And, what about the small 'rabbit hutch'
apartments? And, what of women – don't they need a party to help them
further their career goals?

It is possible that such a middle-class oriented opposition party will
emerge in place of the socialist party, as the workers are absorbed as a
lower middle class, and the middle class of college graduates expands.
Such a party would be less elitist, less patriotic, less samurai-oriented, less
male chauvinist, more individualist, more life-style oriented, and so on.

However, such a party would run into the wall of Japanese tradition.
For, the LDP represents the collective Japanese will – the Japanese volk,
the yamato race, the flag of the rising sun, the samurai sword. Can any
party oppose the cultural cynosure of Japan?

In the USA, the Republican Party 'owns the flag' and represents
'American culture' at its most pristine. However, in the USA there are
millions of diverse immigrants and displaced workers who rallied under
the Democratic Party's banner, and blacks and women, who found a home
there too. In Japan, there is no large excluded group. Though there are

'untouchables' of Japanese and Korean origin, their numbers are too few to form a political movement.

Thus, at the moment, the LDP embodies Japanese culture, economic success, and political stability. No party can rule without the LDP. This certainly could change in the future. A real 'civil society' could, and probably will, emerge.

3 The New Class Structure Engendered by the High-Technology Economy, the Bureaucratic State and the Service Sector

THE NEW CLASSES

The high-technology capitalist industrial system, and the complex bureaucratic state connected with it have given rise to a whole new set of strata which we must analyse in terms of their impact on the emerging political process of the modern world.

The old upper class of capitalist industrialists, for instance, has been transmogrified into a class of high-tech global financiers – not rentiers, not passive stockholders, not semi-passive board-of-directors members. These high-tech financiers have become active capitalists – especially in the USA – intervening directly in the corporate process in terms of takeovers, mergers, and the hiring and firing of management teams.

It is also important that the globalization of the industrial and financial markets has engendered a revival of the entrepreneurial business class, in both the developing and the advanced industrial nations. Pacific Rim, Eastern European, and Latin American businesses are thriving – large and small – as are American and European businesses.

The high technology economy, however, cannot run on business principles alone, but needs also the expert skills of managers and technocrats. The high-tech industrial productive process, after all, is still the heart of the modern economy, even though the financial and business principles still guide the market processes. Therefore, an analysis of the managerial, bureaucratic, and technocratic classes becomes essential.

In Japan, the managers of the huge conglomerates, along with the state bureaucrats, run the economy, with the business and financial strata carefully controlled under their authority. This model – with its class constellation – may catch on in Asia. In the EU there is an equal partnership between the corporate managers and government bureaucrats, with the banking and financial markets carefully controlled.

Thus, in the three models of high-tech capitalism, different upper classes have attained differential amounts of power and influence.

In Britain, the gentry still exert a great deal of authority, while the new corporate-managerial and business strata are only beginning to gain legitimacy and influence. Thatcher tried pugnaciously to empower the business, corporate and technocratic classes, over against the gentry and working class. She succeeded to some extent, and now, Britain must decide whether to go the EU route or the American. British ideology and economic theory pressure that nation toward American-style business-financial dominance. However, British economic realities are pushing Britain into an alliance with the EU. The results are not yet clear.

Along with the new upper classes of big business, stock market financiers, corporate managers, and government bureaucrats, a whole set of middle-class strata has also emerged. The middle managers and bureaucrats, the technocrats, professionals, and service workers make up this new middle class. This, of course, is Mills' 'new middle class'.

Yet, Mills' 'old middle class', of small businessmen, shopkeepers, and farmers, still exists, and in some areas of the world – as in Asia and Eastern Europe – this class is expanding rapidly.

Along with these middle-class strata, a very large lower middle class has also emerged. This class is made up of lower-level white-collar workers, who are expanding, rather than contracting, as office work becomes computerized and technologized. The level of computer and electronic skill demanded of these white-collar workers is rising rapidly, but this stratum is learning these needed skills.

Taken together, the new middle class, old middle class, and lower middle class could become a majority class in all the high-tech societies.

The blue-collar working class is going through a dramatic decline in the advanced industrial nations, and a rapid rise in certain developing nations. The Asian working class is exploding in numbers, while the Latin American may soon expand.

Finally, the new high-tech economy also spawns an underclass of working poor and unemployed lumpen poor. The effect of this class can be profound, for, if too large, it can destabilize the political situation, give rise to violent ethnic conflicts, or drift toward crime, vice and violence. Brazil, Mexico, the Philippines, and the USA, are all feeling the impact of this class.

Let us analyse these strata more closely.

The New Upper Classes

We have described the different forms of high-technology industrial capitalism. In each form, the power and role of the new upper classes is somewhat different. The financiers, for instance, have gained tremendous power
over the economy and the polity in the United States, whereas, in the EU,
the corporate managers and government planners have retained most of the
power and control. In Japan, of course, the government bureaucrats dominate the economy and the polity, though the corporate managers have a
great deal of influence.

Let us look at each of these new strata more closely.

The Financiers as a New High-Tech Business Stratum

With the globalized, computerized financial system engendering enormous
profits for those able to speculate and invest within it, a whole new stratum
has emerged in the economic world. This new stratum is made up of both
individual speculators, financial firms, and investor firms – these latter
both play in and administer the global capital markets. Insurance companies, pension funds, corporate financial branches (such as auto-loan segments of major auto companies), and 'Wall Street' financial firms (whether
they are on Wall Street or in London, Singapore, or Tokyo) – these are the
big players. These companies, and individual investors, are transnational
in reach.

What are the elective affinities that this new financial stratum carries?

First, many individuals and many firms have become fabulously wealthy.
New 'Fuggers' and 'Rothchilds' have emerged world-wide. The wealth
that is generated, however, is not channelled into the industrial-productive
process in any direct way. That is, where the financial stratum predominates, as in the USA, money is sent throughout the global financial network,
with the goal of making more money. If production makes money, the finance capital flows toward it. If it does not, then the capital flows toward
an investment that does make money.

In the USA, this has generated a vast global investment process, some
of which has been quite risky. But, since the US government has been
willing to bail out failed ventures – such as the Mexican in 1995 – the risk
has been lessened.

Further, in the USA, in order to make industrial investments pay more
rapid short-term profits, mergers and downsizing have been initiated. This
reduces capital expenditures in salaries, while increasing the short-term
profit-taking potential.

In Japan and Germany, this kind of financial speculation has been kept to a much lower level, with industrial production still in the hands of the corporate managers and government bureaucrats.

Thus it is in the USA, and also in Britain, that the more purely capitalist style of economics has lent greater power to the new financial stratum. Will the new financiers remain in permanent control of the American and British economies? While the British economy is moving towards the EU style of production – albeit reluctantly – the American economy is still leaning toward the more purely capitalist financial style of operation.

The United States and Japan represent the two outer extremes, as Lipset[1] and Thurow[2] recently suggested, with the EU in between.

Therefore one would expect that the high-tech global financiers – our new 'Fuggers' – will continue to dominate the economy in the USA, and will gain influence in the world global markets.

Furthermore, within Hong Kong, Singapore, Thailand, Taiwan and soon China, this financial stratum is growing and will continue to grow. Asian 'Fuggers' have emerged, and their influence within the Asian capitalist economies could grow. At this moment, the Asian financiers are using some of their capital to invest in, and industrialize, China. This is a good thing. But what of their influence within their own polities as they move towards more democratic forms of government?

Tremendous wealth usually leads to great influence within capitalist democratic societies. Therefore, it should not be surprising to find that the financial firms and industrial financiers have developed an inordinate amount of political power in the USA, Great Britain and Hong Kong. Already books like *Arrogant Capital*,[3] *The Revolt of the Elites*,[4] and *A Journey Through Economic Time*,[5] have been warning the American public of the problematic nature of skewing the economy towards the financial segment of the industrial-capitalist process.

Now, not all of the effects of this growth in financial power are bad. The globalization of the economy will undoubtedly be a good thing for the world in the near future. After all, the rapid industrial development of the Asian nations, Eastern European countries, Brazil and Mexico, is quite exciting.

And, further, the weakening of the bureaucratic-managerial strata and of the whole process of over-bureaucratization, with the concomitant strengthening of the business and financial strata is a good thing politically. For it revitalizes the separation of economic power from political power, and, by encouraging business-style efficiency in the corporate units, it overrides the deadening, hierarchical rigidity of ideal-type bureaucratic organization.

It was, after all, the financiers who broke the tendency towards statist-bureaucratic control of the economy and polity, and who re-organized the giant corporations on the more fluid, flexible, efficient, quick-acting model demanded by the market.

The major problem that emerges from the financial stratum is that they undervalue the industrial-productive segment of the high technology economy. Thus, where financiers are powerful, research, development, science, technology and industrial producer-units have to be actively supported, financed, and perfected by the government, or the national industrial system could decline. The USA is facing this problem now.

One last, and very important problem, generated by the new financial rich. This new stratum tends to accumulate so much capital that personal luxuriating and personal power have re-emerged – as they once before emerged during the 1920s in England, the USA, and Europe. That is, just as Veblen described, conspicuous consumption is back. The yachts are bigger, the cars more luxurious, the jewellery more evident. Ten-thousand-dollar watches have emerged where one-hundred-dollar watches suffice, 40000-dollar cars, where 15 000 suffice, 2000-dollar suits where 200-dollar suits suffice, etc.

This is not consumer leadership – such as buying the first television sets or advanced personal computers. What is occurring is the conspicuous display of luxury items that have no value except that of prestige. That is not productive for the economy, and produces very destabilizing class antagonisms within democratic societies. Already, in the USA, the middle class is angry and confused. They envy the 'life-style of the rich and famous', and they are venting their anger on incumbent politicians, and, of course, inevitably, on the poor, instead of the rich.

Furthermore, rich men, like Steve Forbes and Ross Perot, expanding their fortunes through diversified stock portfolios, have been using their wealth to buy electoral candidacies. And, though they haven't won, they gain instant national recognition by buying media time with their wealth. This mocks the egalitarian citizenship ethic which democracies are based on at their core. New-rich candidates in Italy, France, and Brazil, have emerged along with those in the USA.

One last observation: the new financiers – whether American, British, Brazilian or Chinese, are not inhibited by any puritan religious ethic, as were the early capitalist rich, prior to the 1920s in the USA. Why is this problematical? Because this new high-tech financial stratum tends to flaunt its wealth, as did the business elite of the 1920s. In capitalizing nations, such as China, Korea and Singapore, this could be problematic – especially where there is poverty and the middle classes are weak. Complaints are already being heard in China about the new capitalist rich taking servants and concubines as of old. The Confucian ethic demands 'benevolence' towards the 'common people', yet it never inhibited the 'gentlemen' from flaunting their luxury and domineering over the common people. As democracy emerges, this could become a destabilizing factor. It is already destabilizing in the USA and Brazil, for instance.

Conspicuous consumption and excessive power are serious problems emerging with the high-tech global financial stratum, as a new stratum of super-rich.

The Managers as a New Upper Class

Weber was the first theorist to see that the size and complexity of the modern economy and the modern state would necessitate bureaucratic administration. Weber analysed bureaucracy and the bureaucratic classes it would engender.[6] Government bureaucracy, which had already emerged in Prussia, would become an inescapable component of the modern state, according to Weber, and he warned that this could lead to the decline of parliamentary democracy.

Weber also believed that even though capitalism would prove superior to socialism, (in terms of economic efficiency and productivity),[7] the giant size of the industrial corporation would necessitate bureaucratic administration within the capitalist firm. Weber warned that even within the capitalist system, then, bureaucratic domination could occur.

Following Weber's analysis, James Burnham, in *The Managerial Revolution*,[8] established that the managers were, in truth, beginning to control the economic system, over and above the businessmen who owned the companies. Berle and Means[9] went further with this analysis, showing that indeed ownership and control of the corporations had become separated. The business class, in this analysis, had been relegated to the passive stock ownership role, while the managers had taken over day-to-day control of the corporation.

Now, we have emphasized that the new financial class has taken ownership of much of the corporate stocks, and that they have not been passive owners, but rather have forced mergers and breakups, hired and fired management teams, and directly impacted the industrial structure of the economy.

The point we wish to emphasize here, however, is that whoever comes to 'own' the stock – the financial houses, the general public, the employees or the managers – the corporation will continue to be run by the managers.

Even with this continuing power and influence of the big business and financial stratum, the managers have rapidly become a power class in their own right, challenging the capitalist rich (especially the heirs of the capitalist fortunes, who are often less interested and less qualified for business control than their fathers), and the new financiers, for control of the corporation, and even for ownership of the stocks.

Ever since Djilas[10] wrote his treatise *The New Class*, Marxist scholars have expanded upon 'new class' theory. Djilas was referring to the communist party functionaries, who were acting as a new ruling class in the communist nations, and he used examples from his own Yugoslavian case.

Djilas' 'new class' theory can be compared to the 'managerial class' theory that emerged from capitalist democratic nations and was articulated by scholars such as Burnham and Chomsky, and, of course, C. Wright Mills.

Looking at this new potential ruling class, the managers, the question we are asking is: what would be the political difference if a managerial elite replaced the capitalist rich as the upper class of the newly evolving society?

It has been our contention – and we have written a previous volume[11] on this – that one of the unique hallmarks of the capitalist rich is their commitment to democratic political principles and legal authority. In their origins as an independent economic class, the commitment to the limitation of power was essential, since political power was in the hands of the feudal aristocracy, the Catholic Church, and the kingly state. Furthermore, the separation of powers became reinforced by the separation of the economic sphere from the political sphere (and the separation of the religious sphere from the state, as well, which de-deified the executive office and helped contain it within the limits of rational-legal authority).

The Elective Affinities of the Managerial Class

The emerging bureaucratic or managerial upper class has been termed the 'power elite' by Mills, and the 'new mandarins' by Chomsky. In both cases, the characterization is that of a despotic, or at least, authoritarian, upper class. The analysis in both cases suggests that such an upper class would circumvent and eventually overwhelm legal-democratic political institutions and ideologies.

Why is this new class viewed with suspicion by those who value democracy? Why should managerial and technical skills lead to despotic rather than democratic modes of political action or thought?

The new upper class emerges as an upper class through three sets of skills: technocratic, managerial and 'Byzantine' political skills. Let us look at each of these.

Technocratic Knowledge.   It is important to note that the new technology has become an important element in administration, not just in production. Thus, the administration of the economy and the polity, and the social service sector, now demands familiarity with technological knowledge and equipment, especially the computer-information networks emerging.

One can see a growing trend within modern societies towards the absorption of technocrats into middle management positions, and the rise of ex-technocrats – then trained as managers – into the power elite.

In any case, whether the individual is actually trained as a technocrat, or whether he or she is trained as a bureaucrat (manager), a good deal of technical knowledge may become a prerequisite for any individual attempting to move beyond middle levels of management. In capitalist, post-communist, and Third World societies, knowledge of engineering and computer technology, as well as knowledge of the social psychology of large-scale organizations and 'small groups', are becoming necessary skills for modern administrators.

The technocratic elective affinities of the new power elite creates an anti-democratic bias. The idea of majority rule is replaced by the idea of policy formulation by specialized experts. Technical knowledge becomes more respected than generalized knowledge. And technical knowledge is 'amoral' in character – lacking humanistic or religious ethical prescriptions. The amorality and technical specialization of the knowledge of experts is exactly the kind of knowledge which bureaucratic managers prefer to deal with.

Managerial Skill. If scientific and social-scientific technical knowledge are becoming prerequisites for managerial advancement, administrative and political skills are still the keys to control.

Let me separate administrative ability from political skill for the purpose of clarity. The former has to do with an executive's ability to carry out large-scale productive or service tasks, while the latter has to do with manoeuvring for upward mobility in the hierarchy, domination of subordinates, and elite control of the organization, bureaucracy, or nation.

Administrative ability is linked with leadership qualities within large-scale organizations, because the leader (or leaders) of giant bureaucratic organizations must have the ability to facilitate the carrying out of complex, long-term tasks, to delegate authority to subordinates, to tap the creativity of the technocrats and middle managers, to keep up the morale of the line staff, and, conversely, to cut through the inefficiencies built into the hierarchy, and the regulations, and to reduce the conflicts that often emerge. Administrative or managerial ability is rewarded by modern organizations. Administrative leadership is a special kind of leadership linked to the unique modern rational bureaucratic organization. It had its counterpart in the ancient world among those bureaucratic leaders charged with accomplishing the vast corvée labour projects for which the ancient empires are rightfully famous.

The effective administrative leader – neither charismatic leader nor hack, but definitely not the democratic-plebiscitarian leader – emerges as just that, the behind-the-scenes organizer of large-scale tasks who is able to coordinate an enormous and complexly differentiated staff, and who is able to utilize the authority of the hierarchical chain of command, and able to cut through it when it is unresponsive, or resistant, to the task at hand. The effective administrator must be able to cut through the blockages in the bureaucratic hierarchy and red tape by calling on identified 'natural leaders' beyond those of the designated hierarchical chain of command. He or she must be able to delegate authority to staff leaders as well as 'natural' leaders and to 'get things done' beyond the structure of the rules where these rules become an impediment to the task. Managerial skill also includes the ability to reorganize the structure of authority units and task groups towards the good of greater efficiency and productivity. Managerial leaders who are able to restructure large-scale organizations in an effective way may gain an excellent reputation among their peers.

Byzantine Political Skill. If technical knowledge and managerial ability were all there were to managerial leadership, why would a whole literature of pessimism have emanated from the social scientists since Weber's writing became known? The key to an understanding of Weber's pessimism is the political skill demanded as a prerequisite for entrance into the power elite. By political skill in the bureaucratic system we do not mean, of course, those abilities which characterize plebiscitarian-democratic leaders. We also do not mean those abilities which characterize demagogic-dictatorial leaders. In the context of the modern bureaucratic organization, we mean something like a neo-Byzantine skill for political manoeuvring within the hierarchy of large-scale complex organizations.

Scholars use the term 'Machiavellian' to denote the direct use of naked power by political leaders for their own ends. However, Machiavelli's Prince is not the model for the perfect elite manager. The 'Prince', operating in a complete anarchic vacuum, with all legitimacy and authority destroyed by the inconclusive class, family, and regional struggles, had to resort to the kind of naked force and murderous terror that would be an unnecessary embarrassment to the elite managerial leader.

The managerial elite operates from within a well-structured hierarchy of authority in a political structure wherein new forms of legitimacy (emanating from the administrative organization of society and the new upper class itself) may make the flagrant use of power unnecessary. Hence,

Byzantine, rather than Machiavellian styles of power processes, tend to emerge.

The reason that a neo-Byzantine style of power manoeuvring emerges from giant complex bureaucratic organizations is that such organizations, though rational in their administrative structure, are irrational in their political processes.[12] In fact, they have institutionalized the irrational processes of legitimation. Rational processes, such as citizen's participation, power limitation and law are replaced by elite secret decision-making, unlimited tenure in leadership offices, and administrative rules and regulations. The line officials become 'little dictators', the administrators, 'soulless' bureaucrats, and – with legal-democratic restrictions absent from the bureaucratic organization – the struggle for elite leadership positions takes on the form of complicated, 'ruthless manoeuvring' within the top level of the hierarchy and within the elite circle itself.

If the bureaucratic organization is separated from the power of the state, the power struggles and domination take place without the use of force and without the consequence of violence and death. But the quality or the character of the struggle for leadership – even for the 'executive suite' – is one of neo-Byzantine (veiled) ruthlessness.

The Ideology and Legitimation Process Carried by the New Managerial Upper Class

The new middle class, as we shall show, may very well carry legal rationality as one of its elective affinities, but the new upper class will not. The new upper class's power is fully dependent upon bureaucratic rationality. Legal rationality inhibits its power potential. Whereas legal rationality enhanced the power and legitimacy of the commercial upper class, protecting it from aristocratic kings and popular dictators and allowing it to act oligarchically within a framework of democracy, legal rationality does not protect or enhance the power of the administrative upper class.

The managerial elite has, where legal authority is firmly entrenched, as in the USA, attempted to circumvent it when it conflicted with organizational needs and goals. For instance, boards of education, hospitals, institutions for the care of the retarded, or the insane, corporations, government bureaus, etc. – all of these bureaucratic organizations – through their power elite resisted the law when confronted by legal constraints or legal decisions that conflicted with elite policies. Where legal authority does not exist, the power elite has made little attempt to establish it.

It is significant to note that where, as in ancient China, the bureaucrats were able to create their own legitimizing ideology, that ideology was not one we would call democratic. That is, it was not characterized by the ideals of participation, limitation, and law, but rather by the ideals of social order, deferential obedience to those of higher status, along with an egalitarian recruitment and promotion system (at least as a principle of ideology if not a fact of social reality), and a rigorous, but almost ritualistic, cultural and intellectual education.[13]

This latter mandarin-Confucian ideology gives us a clue as to the kind of ideology which may emerge from modern bureaucratic strata if they come to gain the kind of legitimation they gained in ancient China.

The Bureaucratic State and the Class Structure

The enormous growth of the modern state has been noted and bemoaned by both contemporary citizens and democratic theorists. The larger and more wide-reaching the scope of the state, the greater the threat to democracy. Having just emerged from the era of fascism and Stalinism, few modern individuals are naive about the totalitarian potential of the contemporary state.

But it must be established that against the theory of the minimal state, central to legal-democracy, the modern state has grown dramatically. It has grown in its military, spy and police segments, as well as in its service and regulatory segments.

The Pentagon, for instance, is the largest office building in the world, employing thousands of bureaucrats, white-collar workers, technocrats, and, of course, military personnel. C. Wright Mills, in *The Power Elite*,[14] has described in detail the way in which the military bureaucracy has developed independent power from the civilian government. The military has its own surveillance (spy) bureau, makes advertising and propaganda films, puts on displays of aircraft, ships, and so on. It is hardly controversial that the modern military has become a huge organization, even during times of peace.

Those power-institutions are, of course, frightening, but civic bureaus, such as the tax bureau, the highway department, the Housing Administration – and a thousand others – make decisions that affect our lives every day. And, in fact, they may impinge upon the average citizen more than the military or spy bureaus in most modern nations.

For instance, many years ago, the Atomic Energy Commission in the USA ruled that nuclear power plants were safe. This set off a national, then a world-wide programme of nuclear power. All of us will have to live with the problem of nuclear waste because of this bureaucratic ruling that never came to a congressional debate.

This latter example exhibits perfectly Weber's fear that the knowledge of experts in government bureaus would override the ability of congress or parliament to rule. We shall discuss this problem and some solutions to it later on in the section on controlling bureaucracies.

Finally, the service sector of modern states has also grown dramatically – both within the state and as adjunct to the state. Public schools, public hospitals, national health insurance organizations, family assistance bureaus – these, and hundreds of other services, such as pension, social security and unemployment compensation, are essential services which cannot be cut and probably will be expanded.

The services, in themselves, are good – most citizens want them, and, in fact, can't survive without them. But the huge bureaucracies that administer these services are politically dangerous. They are despotic structurally, and, by oversight or intent, sometimes destroy the lives of innocent individuals.

Services like social security and public schools must be guaranteed. But the bureaucracies delivering such services must be controlled and their powers limited. If the bureaucracy is connected to the state, the power situation is worsened, for there is less recourse through the courts against them. But, even where the bureaucracies are private, they can impinge upon and destroy the very people they are supposed to help. The nursing-home scandals in the USA showed this latter problem, for, though they were private, they ran roughshod over the civil liberties of the individuals in their care.

Thus, private or public, service bureaucracies will have to be limited in power if democracy is to survive. Later, we discuss the ombudsman system which the Swedes, Germans and others have instituted to protect citizens from military, civilian, public and private bureaucracies.

Along with the corporate managers, then, a whole stratum of government and service administrators has arisen. Because this stratum is not connected to the capitalist profit system, these administrators tend to exhibit more purely bureaucratic characteristics. They are more officious, that is, have more of the character of 'little dictators'. They may be obstructive, since no profit or production goals press on them. They may possess life tenure, always a dangerous situation for individuals in power positions.

The government and service bureaucrats do tend to dominate their hierarchies like a 'power elite', and, they do tend to act like 'modern mandarins'. We shall fully discuss the political implications of this – the threat to democracy, and programmes for containing this threat.

Let us conclude by saying that the growth of the giant complex bureaucratic state, service sector and corporation have engendered a growing and powerful new class in modern society – the bureaucratic managers. These managers could become the power-limited administrators of a world of abundance, or the power elite of a world without liberty. While the collapse of despotic communism gives us hope in this regard, specific mechanisms of control limiting the power and tenure of bureaucrats and constraining their actions within a framework of rational law, must be developed, or we shall lose this newfound hope.

THE DIFFERENT STRATA WITHIN THE NEW MIDDLE CLASS AND THE VARIATIONS IN THEIR POLITICAL ORIENTATION

The Technocrats

As we have mentioned, Veblen[15] first identified the technocrats ('engineers') as a critical new stratum in the industrial productive process. The technocrats have become the key class in the development of applied technology in the new automated, robotized, computerized industrial productive process, and the computerized global financial market.

Yet, though they are the key stratum of the modern world, like the peasants of old, they lack power, control, and even prestige. Veblen recommended that the 'engineers' should run the factories. But, just as there was not to be 'workers' control' of industry, there has not been technocratic workers' control. As we have emphasized, the managers, bureaucrats and the financiers (old-line or *nouveau*) have come to control the modern economy.

Are the 'engineers' resentful? In some cases – but they are well-paid professional workers, living a middle to upper middle-class life style. They, therefore, may desire more input, more control over corporate or government decisions, but they are hardly a rebellious class. Rather, they are stable, even conservative members, of the new middle class – conservative because they are part of the corporate or governmental organizations and share their general orientation to the world, and, middle class because they make moderately good salaries.

Who are these technocrats? They are scientists, including physicists, mathematicians, chemists, biologists, engineers, computer programmers and others who work in varying degrees of applied or pure research. Some corporations fund 'think tanks' – like Bell Laboratories – where scientists can pursue many areas of pure research. Most corporations employ scientists in a more directly applied capacity, linking their skills directly to their product line or their production process. Universities, however, also run think tanks, and these are often well funded.

When money was available in the United States, think tanks and 're-search and development' facilities were heavily funded. As money has dried up in the United States (for all but military purposes) research and development has declined dangerously. However, as the USA cut back its programmes, Germany, the EU, and Japan have stepped up their pro-grammes, so that now they lead the United States in non-military patents used for industry. Therefore, as the modern economy develops, both ab-stract R&D and directly applied science continue to be exhibited as critical elements.

For instance, thousands of biologists are employed by cosmetics firms worldwide, while hundreds of thousands of chemists are employed by rubber, chemical, plastics, munitions and space industries, and so on, while engineers and computer experts of varying varieties are needed in every kind of modern productive or construction process.

Along with the mathematicians, physicists, chemists, biologists and engineers, new kinds of technocrats keep emerging. There are computer specialists, genetic experts, systems analysts, etc., branching out from the science and engineering fields, and such specialization should continue as the technological revolution explodes further.

These scientists and engineers will become increasingly important as 'knowledge experts', and their status may rise as their key role becomes more fully understood by the population at large. At the moment, though, they remain a middle stratum, employed and controlled by the middle and elite managers of the corporations, government bureaucracies, and univer-sities with which they are affiliated.

The technocrats do become members of the management hierarchy in some cases. And, often attain top management positions. In this case they accede to power – but not as technocrats, rather as managers. In fact, those technocrats who are the most scientifically talented are less likely to move into management positions. Therefore, as technocrats, they remain under the control of the elite managers – although their technical input obviously affects the decision-making process. Therefore, administrative, rather than technocratic skills and talent determine one's rise to managerial power.

Will the technocrats become a future power class? Do the technocrats carry democratic or authoritarian elective affinities?

At this moment in history, the technocrats – scientists, mathematicians, engineers, computer programmers and so on – tend to act submissively within the organization they work for. They are dependent upon the or-ganization for their equipment, research associates, and, of course, for their salary. Very few of them can become independent researchers at a university, almost none can work on their own. The days of Edison are over. Equipment is too expensive, research too narrow and specialized for independent scientific creativity. Therefore, the technocrats have been fully co-opted and usually feel strongly identified with the organization. The work-world 9–5 loyalty of the technocrats is closely scrutinized. Those who are independent of, or disagree with, management decisions will be pushed out. However, the public activities of such individuals, as well as the private activities, are ignored, unless they fly directly in the face of that particular organization's policies.

The technocrats, therefore, have at times been publicly active politically – even in causes that are unpopular with their organizations (as long as their organization is not directly involved or confronted). Scientists, mathematicians, engineers, and other technicians have been leaders and activists in environmental and social protest movements, active in local politics, and in national campaigns. The organizations for whom they work – private or public – do not encourage such political actions, but they have not been punitive in most cases.

Given the dependency of the modern bureaucratic organization on the expertise of the technocrats, and, given the technocrats' exclusion from the power structure of these organizations (except where they become managers), the technocrats, as a new stratum of the new middle class, could become one portion of an active public citizenry in a high-tech democracy, or they may remain anonymous functionaries, submerged in their corporate or government bureaucracies.

Service Workers

The service workers are another rapidly emerging stratum of the new middle class. Psychologists, social workers, doctors, nurses, and other health-care professionals, make up a portion of the service stratum, while poverty lawyers, civil-liberties lawyers and other non-corporate lawyers (corporate lawyers are attached to management as specialized experts) make up another portion of this stratum. Teachers, day-care, nursery, and child-care workers are another segment of the service stratum.

There are many new 'service' areas emerging in mass society where the function of the family, religion, and community have disintegrated, and wherein the isolated individual and the giant organization must turn to specialized experts for any and every kind of help.

It could be predicted that the social service workers will eventually be fully co-opted into the new technological-bureaucratic system and simply work to help people adjust to it. It is also possible, however, that both the humanistic tradition of concern for the individual and the process of heightened self and cultural awareness which are integral attributes of the modern 'helping professionals' will have a liberating impact upon individuals as individuals. Both trends are now occurring – which will predominate depends on other structural and political conditions which will emerge.

The case of the social psychologists is illustrative here. Many large-scale organizations employ – on the payroll or as consultants – psychologists and social workers. These workers run special weekend programmes, workshops and retreats. The results of such 'sensitivity' programmes often transcend that which the organization intended.

That is, the organization wishes to improve group processes within a department or division. The hope is that if the individuals concerned spend an intense weekend (or set of workshop days) together, wherein role barriers and authority hierarchies are temporarily suspended, the group, because they become more intimate and more open with one another, will work more smoothly, cooperatively and caringly together and exhibit less conflict, less blocking; will utilize 'natural' leaders more effectively, and so on. The 'Organizational Development' specialists seem to have a good deal of success with this approach, and it is becoming typical for corporations and government bureaus to utilize small group experts in this way.

However, personal self-awareness and awareness of group process results from these activities. Many individuals subsequently continue this activity by joining consciousness-raising groups, therapy groups, counselling groups, independent of their organization. This may simply produce a heightened narcissism and more of an attempt at private self-development. It could also distance the individual from the organizational structure through a heightened insight into group process and power relationships. Whether this distancing becomes a basis for independent political thought and action is not clear.

As to the service workers themselves, some will become advocates for the people they serve, some will become co-opted into the organizations that employ them. It is not absurd to predict, however, that a kind of humanism is built into the service professionals such that a certain percentage of such professionals – unless repressed by government or corporate power – will focus upon individual liberation, rather than adjustment, and will act as peoples' advocates, rather than the organizations' apologists.

The Intellectuals

One of the 'unintended effects' of mass college education is the expansion of the number of intellectuals produced. A certain small percentage of individuals become enthralled with ideas themselves, rather than with the careers that such ideas might lead to. Whether these individuals incline towards philosophy, theology, literature, art, music, physics, chemistry, or biology, sociology, psychology, economics, etc., is immaterial. What is crucial is that they pursue these ideas independently (or in cooperation with other independent intellectuals).

Now in most modern nations – democratic or despotic – the utilization, co-optation or repression of such intellectuals is actively undertaken by governmental or private power groupings. The talent, skill, and occasional creative genius of such intellectuals is recognized as both a potential contribution to, and a potential threat to, political stability and authority.

Contemporary societies – both democratic and despotic – have succeeded in utilizing and co-opting large segments of the intellectual stratum of the new middle class. However, many of the intellectuals continue to remain independent-minded. Given this, the repression of such intellectuals also becomes a fact of modern society. For independent intellectuals also become critical of the political and social system in which they live, and, their criticism, since it is usually literate and often powerful, may strike a responsive chord in many segments of the society that might be discontented.

The French intellectuals have played this kind of contributive and destructive role. From the *philosophes* to the existentialists to the post-modernists, they dominated the ideological scene in France, inspiring both anti-government and pro-government movements. So too did the German intellectuals – from Kant to Hegel, Fichte to Nietzsche, Weber to Heidegger – generate and reinforce anti-democratic and pro-democratic movements. The intellectuals within the Third World nations have also exerted great and diverse influence there. In communist, fascist, and military-dominated nations, the intellectuals are openly repressed if they speak, write, or act against the government. In parliamentary democracies, repression against the intellectuals is more subtle, less violent, but often quite successful. In the US, during the McCarthy era of the 1950s, left-wing intellectuals were successfully removed from government, university and mass media positions. And, during the late 1960s and 1970s, the left-wing, counterculture, and black intellectuals were harassed by government agencies, university administrators, and the mass media. Right-wing intellectuals, though not harassed by government agencies, have been harassed by university administrations, and have in some instances been stigmatized by television in a quite consciously repressive way.

Why are the intellectuals problematical, and do they exist as a stratum which necessarily fosters democracy?

The intellectuals are problematical simply because: (1) they are free thinkers and will tend to be critical, focusing on the unsolved social and political problems of their societies, and (2) because they know they are very intelligent and creative, yet they are usually not included in the key decision-making processes of the societies in which they exist.

The independent creative thinking of the intellectuals is, of course, the hallmark of the 'civilizational progress' of the human species. Creative ideas – whether they are turned towards economic, political, or spiritual 'survival' – are indispensable to the existence of human beings as human beings. The intellectuals and their innovative ideas do not necessarily contribute to progress in the human condition. They do not necessarily make us happier, healthier, or freer – nor do they guarantee our survival. Such ideas are, in fact, just as likely to do the opposite. However, civilization progress is progress at least in terms of: the technological control over nature, the scientific understanding of the universe, and the social-scientific understanding of human society and the human individual. Power groups, therefore, usually see the advantage of utilizing the intellectuals and their creativity for their own ends.

However, they cannot buy off, co-opt, or even repress all of the intellectuals. Those outside of the official status structure may become critical of the society in which they live – critical because they are perceptive, creative or humanistic, or critical because they are left out of the power and status system.

The humanistic intellectuals will point out to their society those groups which are downtrodden and those groups which are usurpative, along with those institutions which are repressive or unjust; from the Jewish Prophets to contemporary social critics, humanistic intellectuals have performed this role.

The criticism of society generated by the intellectuals need not be engendered by humanistic motivations, of course. From Plato to contemporary elitists, intellectuals have been critical of society because of its lack of order, its lack of planning, and its lack of reliance upon the intellectual elite. These elitist intellectuals, as well as the humanistic intellectuals, will challenge the power structure and point out the shortcomings of society.

The intellectuals, as a stratum, undoubtedly possess a 'hidden agenda'. That is, having been identified in college and university as the most intelligent individuals in society, they come to believe that they probably could rule more rationally, innovatively and successfully than the non-intellectual leadership (no matter what system of government prevails in the particular case in question). Believing this, and being critical by training, the intellectuals become a very threatening stratum within any political system.

Though the intellectuals may not be democratic or humanistic in their ideology, their critical role, along with their opposition to power and status groups, fits in with the definition of the 'concerned citizen' in democratic political systems. The intellectuals are, thus, at once, the backbone of any democratic process in that they assert their free speech, are likely to engage in public debate, and will highlight existing social and political problems, while at the same time, paradoxically, they may not espouse democratic ideas at all, tending toward elitist and authoritarian utopianism consistent with their conceptions of themselves as a stratum of superior individuals who could run things best if given the power.

It should be asserted strongly that a properly functioning democratic system should be able to tolerate and absorb a good deal of criticism and activism from its intellectuals, as long as the system of legal authority is deeply ingrained.

Lower-Level White-Collar Workers

At the bottom of the private and public bureaucratic hierarchies are the white-collar workers – clerks, typists, secretaries, receptionists, telephone operators, etc. Again, the modern economy has engendered an ever-growing need for such workers. With the introduction of the computer, many analysts believed that a reduction of the white-collar work force could be achieved. However, this has not happened. In fact, the word-processing, xerox, and fax machines seem to have expanded the need for paper and have demanded more, and better trained, white-collar staff.

This white-collar work force is beginning to replace blue-collar labour as the majority stratum of the lower middle class. Many of these white-collar jobs were initially filled by women (who were paid less and treated like wives around the office). However, the white-collar jobs are now unisex, as are the professional new middle-class positions (although gender discrimination continues to exist, more so in Japan and Europe than in the United States, but in every industrial nation).

The lower-level white-collar workers are underpaid, and have little authority within the hierarchies. Therefore, they have tended to unionize (as have other new middle-class strata), in order to gain better wages and working conditions. Whether the white-collar unions will become militant (they have in Sweden) as did the blue-collar unions, it is too soon to tell. However, such unions could play an important role in the future society. They also may not be critical, if the salaries and living standard of the white-collar class are improved commensurately.

The lower-level white-collar workers are often underpaid and overworked.

Office work can be high-pressure work. The pressure put on the middle managers puts pressure on the white-collar employees. Overwork, pressure, and underpayment can produce anger in the lower-level office workers, who do not share in the rewards the middle managers and technocrats may gain. Therefore, unionization has gained strength among the white-collar employees.

There is great variation here. Within some corporations and government bureaus, the white-collar employees are well integrated into the total organization. They feel loyal, part of the team, and are willing to work hard for the company or bureau. In other large-scale organizations – government and corporate – there is antagonism, discontent and militant unionism.

Large-scale organizations can co-opt white-collar workers with job security, fringe benefits, decent wages and personnel policies geared to morale boosting and cohesiveness. However, they must be prosperous and/or forward thinking in order to accomplish the latter. Thus, the white-collar stratum may become a lower middle class, or it may become a new proletariat.

If the lower white-collar workers become proletarianized, then white-collar unionism may play an important role in the political processes of modern societies. Some theorists, such as Lipset,[16] suggest that this kind of 'adversarial' process is healthy for modern democracy, while others, such as Dahl,[17] suggest models for processes of industrial democracy which focus on participation by white-collar employees, rather than union adversarialism.

Small Business (the Old Middle Class)

We must be clear on the fact that small business is declining as a class. The malls, the franchise stores, the national chains – they have all made it difficult to compete in the small business market. And most of the retail business has been converted to the big-business managerial–new middle class model. This is happening in Europe and Japan as well as in the USA.

However, small business opportunities, in the areas of boutiques, restaurants, cafés and other speciality shops, continue to persist. I mention this because the small-business middle class has not disappeared. Both indigenous and immigrant groups in the advanced industrial societies, continue to utilize small business as both an avenue of upward mobility and a way of life.

The continuing existence of small business – though often conservatizing in social values – is democratizing in political culture. Mills' characterization of the independent attitude of the old middle class still holds. In both Europe and the United States the small business class is politically active and inclined toward democratic values. We must not forget the fascist interlude, however. Under certain conditions the conservative social values of the small business class can certainly by turned toward anti-ethnic or otherwise fascist actions and ideologies. Given the existence of a growing underclass, tending toward crime and violence, the fascistic attitudes of the small business class (and the entire middle class) could be played upon. The prosperity generated by the new high-technology economy, however, is so overwhelming that barring a major depression, it is doubtful that such a fascistic trend could be generated – though this is always possible.

I wish to emphasize the positive side of small business for a moment, though. For in Eastern Europe, Russia, and China, a small-business class is beginning to emerge, and the effect of this class may be very good in terms of democracy. Since the new middle class is very large and well-educated in the post-communist societies, its expansion due to the addition of small business could help to create the middle-class majority necessary to democratic stability. Many workers and peasants are using small businesses as a method of upward mobility. Their success and their link to market production bodes well for the potential of legal democracy in the post-communist nations.

The problem, however, of some small-business persons becoming rich in the post-communist societies is something that may be difficult for the people to handle, given their past ideology. We have to wait and watch this phenomenon with great caution.

The old middle class of small-business persons and farmers will never have the impact they had during the nineteenth and early twentieth centuries. However, the continuing existence of small-business opportunities helps swell the ranks of the middle class in the advanced industrial societies.

Conclusions on the New Middle Class

Since C. Wright Mills wrote *White Collar*[18] social scientists have been aware of the emergence of the new middle class. And Mills made us painfully aware of the difference in character and professional circumstances between the independent, entrepreneurial old middle class and the salaried, conformist, bureaucratically entrapped new middle class.

The small-business middle class – the old middle class – has not disappeared, but they certainly have declined in relative numbers and influence. The new middle class has become the majority segment of the vast and growing middle strata spawned by the new mode of production. The question we must turn to, then, is: will the new middle class become an

Aristotelian middle class – that is, a middle class carrying legal-democratic values and processes, and able to act as a mediating class between the rich and the poor?

C. Wright Mills was pessimistic on this, believing the new middle class would become locked into the status hierarchy of one giant bureaucracy or another, and that they would not be able to exert any independent, democratic influence on the society at large. Mills believed that the managerial power elite of the corporations, the government, and the military would become a despotic elite, slowly but surely eroding parliamentary democracy, and relegating it to a secondary level of power.

Max Weber, writing before Mills, attempted to warn us of the 'iron cage' of bureaucracy closing around us. Weber painted a picture of a future world run by bureaucrats and technocrats 'without soul' from which we would withdraw into a world of private sensuality 'without heart'.[19] In his most pessimistic moments, Weber predicted that the world of modernity would become like that of the ancient empires – that we would end up like the Egyptian fellaheen: well fed, well clothed, but stripped of all political rights and lacking all moral dignity.[20]

I call this the 'Empire Analogy', as opposed to the 'Polis Analogy', for the modern world. Neither Weber nor Mills wanted the Empire Analogy to emerge. Weber especially was urging us to prevent the Empire Analogy from occurring. He wanted us to save parliamentary democracy from bureaucratic authority and technocratic expertise, and to retain our public citizen's role. Can we break the bars of the 'iron cage' of bureaucracy? Can we retain and vitalize democracy in the age of the new middle class and the managerial upper stratum? These are critical questions, the answers to which will determine our near-future political fate.

Balancing the heavy pessimism of Weber and Mills, the recent events in Eastern Europe have generated a new optimism about the possibilities for democracy in the modern world, and the limits of bureaucracy. That is, in the Eastern European nations, including the Soviet Union, the emergence of a growing, well-educated new middle class generated an overwhelming drive toward democracy, the protection of law, and the desire for a consumer-oriented economy. This process is occurring in China as well.[21]

Few analysts believed that a democracy movement was possible, much less probable, wherein the population was locked into the rigid authoritarian, dictatorial-bureaucratic Stalinist-style communist state. Yet, with the production of a large, well-educated middle class, and a working class moving up to a lower-middle-class status, an explosion of democratic idealism and consumerist realism overwhelmed what appeared to be entrenched despotic-bureaucratic regimes.

However, the warnings of great minds must never be ignored. And, the Empire Analogy hangs over us, and could envelop us in its bureaucratic-technocratic cloak. Our 'Brave New World'[22] could become a world of abundance without freedom – a yuppy world of affluence, hard work, private self-actualization and sensuality, political dependence, and sociopathic amorality – the culture of narcissism.[23] It could become something even worse – a 1984[24] totalitarian world, with the high-technology control capacity beyond even what the Nazis possessed.

However, the new middle class, in alliance with the working class, educated to the rational-scientific world view, disgusted by the inefficiency of the bureaucrats, suspicious of the decisions made by 'experts' (who seem so often to disagree or simply err) and painfully aware of the growing power of the modern state, may yet become an Aristotelian middle class, shoring up democracy on a high-technology base.

Will we see 'television town meetings' on cable-computer systems, or will 'Big Brother be watching us'?

But before we discuss such a project, let us further analyse the class structure. For the position of the working class has changed, and finally, and most unfortunately, an underclass has been inadvertently generated by the educational pre-condition for the high technology work force. Let us analyse each of these.

THE NEW LOWER CLASSES

The Decline of the Industrial Working Class and Its Ramifications

The industrial working class, in the nations which have developed their high-technology economies, will decline in numbers and importance. Automation – or machine production tended by technological devices rather than workers – will eliminate the need for a massive blue-collar work force. However, we have already mentioned that at the same time as the work force is declining in the advanced industrial countries, it is expanding in the Third World.

This is not paradoxical. It simply reflects the fact that certain products – like clothing and sneakers – still require blue-collar labour, and, certain processes in the high-technology production system still require labour – like assembling and packing electronic equipment. Therefore, in their typically capitalist quest for cheap, non-union labour to exploit, the high-technology firms themselves – be they American, Japanese or the EU countries – have moved this portion of their production operations to the

Third World. Therefore, while blue-collar labour is declining in the advanced industrial nations, it is expanding in the less developed ones.

The effect of joint-venture capitalism on the Third World is complex. The positive side is the creation of jobs – albeit low-paying jobs, but still jobs, which in a poor nation are a good thing, exploitation or not – and the beginnings of a consumer goods economy with an indigenous new middle class emerging to produce, distribute, and consume the high-technology products.

On the negative side, the exploitation of the workers can be Dickensian, and worse, those lured to the urban centres for work, but not finding it, often end up living the life of a lumpen proletariat in the hideous squalor of the slum-barrios of the Third World. Nor can indigenous industries compete with the high-technology companies invading their shores.

But let us analyse the decline of the working class in the advanced nations. Automation, robotization, and computerization have combined to reduce the need for an industrial labour force. As with the Third World, this has produced both positive and negative effects. The positive effect is that the children of the working class are going to colleges and universities and they are moving into the white-collar, professional, and technical job world. This is genuine upward mobility, and should be viewed in the positive light that it connotes.

The children of the working class, of course, are not usually as good students as middle-class children – their background is less intellectually enriched, and their families tend to think in rigid moral and religious categories or clichéd analogies. Therefore, the children of the working class tend not to do as well in college; nor do they test well enough to gain entrance into the elite colleges; nor can their families afford to send them to the elite colleges if they do test well.

However, the state universities in the United States, the 'redbrick' universities in Britain, the gymnasia and universities in general in Europe, are very good institutions. And, the children of the working class do well enough therein to qualify for and gain employment in the white-collar, service, middle management, and professional career tracks. Even though in most cases they do not do as well as the middle to upper middle-class children, they do well enough to become stably absorbed into the new middle class.

Now, the children of the working class do experience terrible pressure, during their teen years, in terms of school success. And before we speak glibly of upward mobility, we should remember the high rate of teen suicide emerging in the USA and the increasing drug and alcohol abuse among teens in all of the industrial countries, and the skinhead and hippie withdrawals from the culture. There are other causes for these youth problems that are equally important, but school pressure and the inability to compete academically with the middle-class students, do cause anger, violence and deviance. These problems emerge in Britain, Germany and even Japan.

Worse than the problem for the upwardly mobile working-class youth is the problem of the downwardly mobile worker – the worker who loses his or her job because of automation, and cannot upgrade his or her skills due to lack of educational background. A horrific negative effect of automation has been the rising unemployment rate among previously unionized workers who had steady jobs. This 'structural unemployment' was first noted in the 1960s, but now has become a mass phenomenon. All over the industrial world, but most conspicuously in the American 'rust belt' around the Great Lakes region, and in the old factory districts of Britain, especially Manchester and Liverpool and Edinburgh, unemployment due to automation – or the corporations moving the industrial operation to the Third World – has risen to dangerous proportions.

In Germany and Japan the corporations and the government are dealing with this problem very well. Excellent training and upgrade-educational programmes have been developed to try to bring workers up to the standards and skills necessary for technical jobs. But in Eastern Germany the unemployment problem has grown dangerous. Many German (and Japanese) workers successfully make the transition to the high-technology white-collar world, East Germany not withstanding.

Job-placement agencies exist, wherein unions, government and corporations cooperate to help displaced workers find new jobs, where such jobs are available. Finally, where jobs are simply unavailable, or if workers cannot be retrained, the Germans and Japanese still have an option: the Germans send the workers home, for the majority of their work force are guest-workers from Greece, Italy, Yugoslavia, Turkey, etc. (although the Turks usually will not go home, and, of course, the East Germans are 'home'), while the Japanese keep the unemployed workers on payroll and put them to work at beautification and maintenance work! This latter may be a cultural phenomenon, but is an interesting transitional alternative.

Needless to say, in the United States the corporations and the government do nothing for those workers automated out of a job. We talk about job re-training programmes, debate a national job placement process (the job-bank proposal), but basically abandon the unemployed workers. Of course, they get unemployment compensation for a fair period of time, but they often get little after that.

What happens to American workers who cannot find work? The record is frightening: from among unemployed workers (who had once held decent-paying union jobs) a rising rate of murder and family violence is emerging. Everyday in the United States, one reads of a laid-off worker returning to his plant and shooting the supervisor who fired him and anyone else happening to be in his way, then turning the gun on himself. Sometimes the shootings are more random, occurring at a mall or a McDonald's. Often the violence occurs at home – unemployed men taking their anger out on their wives and children. Since unemployed women are more likely to find white-collar work than unemployed men, and since working-class men tend to be ultra-macho, and in America, hunters and gun-owners, the violence tends to emanate from the unemployed men. This kind of violence is beginning to spread in Britain, Germany, and, of course, most frighteningly in Russia, where massive unemployment has occurred during the 'shock'-transition to capitalism.

Thus, though the children of the working class may be attaining mobility into the new middle class, the workers themselves are suffering a terrible period of attrition and insecurity.

The Decline of the Political Power of the Working Class and the End of the 'Inherent Leftward Drift' of Industrial-Capitalist Societies

Within the industrial-capitalist nations, the working class, through its union organizations and through participation in the social democratic parties, gained political power, and with it, economic upward mobility. From 1890 to 1940 the workers gained better wages, better working conditions, and a more middle-class life-style.

Of course, the Great Depression interrupted this process and threw many of the workers into a condition of unemployment and economic hardship. However, the depression also weakened the capitalists, and brought the socialists to power in certain nations. In Scandinavia, the Social Democrats took office, and remained in office through the post-war period. And in Great Britain the Labour Party finally took control of Parliament. In the United States, of course, Roosevelt and the New Deal swept into power, and for twenty years instituted programmes not unlike those of Scandinavia. We have discussed fascism elsewhere[25] but it should be remembered that the fascists of Italy and Germany absorbed some of the socialist programmes in order to gain working-class support.

The point is, that from 1890 to 1940, and then from 1945 to 1965 there was a kind of 'leftward drift' within the industrial-capitalist societies. This leftward drift was accomplished by the pressure from the unions and the socialists and the concomitant weakening of the capitalists by the depression and then the war (in European nations).

From 1965 onward, however, capitalism became greatly strengthened in Europe. In the USA the trend occurred in the 1940s and 1950s – the war strengthened capitalist industries, and its aftermath strengthened US industries even more – with no competition yet from Europe or Japan. Even in the USA, though, the unions and the New Deal wing of the Democratic Party were still strong. So too were the unions and the socialist parties in Europe.

The Communist scare of the 1950s and the subsequent repression of the American Left began to destroy the New Deal tradition in the United States, though it did not weaken the European Left.

The real alteration, however, was not the political climate of anti-communism, but rather the advent of the new mode of production. As industrial capitalism transformed into its most modern form – high-technology industrial capitalism – the need for blue-collar labour declined. With automation and computerization, the factory labour force will continue to decline.

This is good in the long run, as mentioned, since the children of the working class will be absorbed into white-collar and technocratic work. However, in the short run there will be continuing unemployment amongst the workers who cannot be retrained, and the blue-collar unions and political parties will lose their base of power.

This is occurring in the USA and in Europe as well. What I wish to emphasize here is that with the decline of the working class and their political power, a conservative trend has set in. The United States and Europe are driving to the right, not the left. This is problematical because the pressure from the left forced redistribution of capitalist wealth from the upper class to the workers and the middle class. The Europeans have extensive redistribution mechanisms, and so does the United States. Progressive taxation, social security and welfare state programmes and services all served to prevent the industrial-capitalist societies from skewing toward the rich and the poor. The middle-class majority itself was maintained by the left pressure.

With the left pressure declining, and the conservative trend gaining (and made more conservative by the collapse of communism – which is a good thing in itself, but may have negative consequences in the capitalist countries), redistribution of wealth from the rich to the workers and the middle class may be stopped. Already in the United States we have seen this trend. Kevin Phillips in his book *The Politics of the Rich and Poor*,[26] documents this problematical situation, and we are also beginning to hear grumblings in Germany and Japan about this skewing of the wealth.

The Marxian prediction that capitalism tended toward the selfish accumulation of wealth by the rich, and their refusal to share it with the workers, was only made false by the organization of the workers into unions and parties, and the political pressure from them to force a redistribution of wealth and services.

What will force such a redistribution now? Will the white-collar unions be militant enough to do so? Will the new middle class demand their share as a class, or just selfishly cull their salaries and perks as individuals? (Yuppyies aren't radicals – they are the new conservatives.) Will the capitalists, financiers, and managers selfishly loot the system, like the big businessmen of the roaring twenties? All of this is unclear. However, what is clear is that the pressure from the left for an equitable redistribution of income has declined with the declining working class, and its weakened unions, and weakening socialist parties. Now, let us make the situation worse.

The Growth of an Underclass

We have emphasized that the new high-technology mode of production demands a college level of education in its work force. We have also emphasized that this is a good thing, because an educated citizenry has the best potential for being a democratic citizenry.

However, what if an individual cannot attain a college education? Then, the job market is bleak indeed. With thousands more workers being laid off each year, due to automation or relocations, good-paying blue-collar jobs are declining. Therefore, for the uneducated, there is little chance for steady work, much less upward mobility. In a nation like the United States, which has always encouraged in-migration from poor countries, the decline of the blue-collar job market is a major disaster. This nation is still opening its doors to the downtrodden of the world – this is what makes it great: the 'melting pot' tradition should be maintained. But the US must also be aware, as a nation, that the economic system has changed. A comparison of ethnic immigrant groups will make the problem clear.

During the last fifteen years, immigrants from Korea, Greece and Latin America have come to the United States. The Koreans, Greeks and Latinos all opened up small businesses – the Koreans opened fruit and fish stores, the Greeks luncheonettes, and the Latinos bodegas (or grocery stores). But the Koreans and Greeks value education, and they understand the need for a college education in the modern job market. They have, therefore, coached and chaperoned their children through the grade school systems and into the colleges – and therefore, the Korean and Greek children have moved rapidly into the new middle class.

The Latinos come largely from rural villages or urban slums. They come seeking labouring jobs like the European peasants of the nineteenth and early twentieth century who came to America. They have no educational background and simply lack the ability to coach their children through school. They do not see the link between college education and the job market, and continue to seek blue-collar jobs. Furthermore, they do not value education, in and of itself, the way the Koreans and Greeks do. None of this is meant as a critique of the Latinos. We only wish to establish that if a group does not have the education and skills demanded by the new mode of production, their economic condition will decline.

The Latinos do work hard – but they cannot help their children through school; therefore, they must take all the low-paying jobs that still exist in modern society – everything from supermarket to fast food to domestic work – all the poorly paying, scut-work jobs. And, either they become members of the working poor – the working class that is non-unionized and paid too little to really live on – or they drop out of the labour force and into the underclass.

In Britain, a similar immigrant situation has emerged. That is, whereas Indian immigrants usually move rapidly into the middle class through retail and professional positions and educational success, most West Indians often drift into the underclass. Pakistanis and Bengalis have gone both ways.

France and Germany may experience these same problems as immigrants from Africa, the Middle East and Eastern Europe flood into those nations.

The underclass expands as union labouring jobs contract, in nations where a sizeable non-educated population exists or arrives. The problem of the underclass becomes more than a problem of unemployment. Faced with a world of displayed consumer abundance and a lifestyle that is dazzling, the underclass – blocked in legitimate channels of upward mobility – turns to crime, vice and violence.[27]

The American drug-dealer, driving his pimp-trimmed BMW, wearing his double-breasted $500 suit and tie, and sporting his automatic weapon, is one symbol of the underclass and its desire to attain the yuppy lifestyle of the new middle class. The crack addict, murdering the local store owner in a desperate robbery attempt to gain money for more drugs, is another. The high-technology economy cannot absorb the underclass, therefore they collect in huge slums, and turn toward illegitimate and violent methods for survival. Thus, what would have been a potential working class in a system of industrial capitalism, becomes an underclass in a system of high-technology industrial capitalism.

In the United States, wherein a mass migration of Southern rural blacks moved to the northern industrial cities, and a mass in-migration from Latin America followed that move, the underclass problem is grave indeed.

But the Europeans are beginning to worry about this problem as well. For Turks, Arabs, Africans, Indians, Pakistanis, South East Asians, Eastern Europeans and others are beginning to accumulate in European cities. Many of these new immigrants will fail to make the educational jump necessary for success in the high-tech economy, and therefore an underclass may appear in those nations. Brazil and the Philippines have also spawned massive underclasses, as they have begun their industrialization.

The presence of an underclass living in poverty and engaged in violent crime can radicalize or conservatize the new middle-class majority. Under certain political conditions they can be moved to humanistic compassion toward the underclass and support (even participate in) programmes for their uplifting, while under differing political circumstances the new middle class can become callous and hostile toward the underclass, desiring their imprisonment and punishment, and even their expulsion. We have witnessed both sets of attitudes – the sympathetic in the 1960s and the rejecting in the 1980s. This flip-flopping had occurred in Europe as well, with the hostile attitude outweighing the sympathetic, in the US in the 1990s.

Whatever attitude the electorate may take, the underclass is a structural result emanating from the educational demands of the new economic system, and it becomes a political and social problem which must be attended to in one way or another – either through rehabilitation or repression.

CONCLUSIONS

Maintaining a proper class balance within the industrial capitalist system – and now, within the high-technology industrial capitalist system – is absolutely essential for the stability of the legal-representative-democratic system. For, if the rich become too powerful, they take the wealth selfishly to themselves. And, if the poor become too poor they tend toward criminal violence and revolutionary extremes. While, if the middle class is weakened, either in numbers or prosperity, 'extreme' democracy of the poor and actual oligarchy of the rich could threaten the legal-democratic system itself.

Thus, the continuing existence of the big business class and the capitalist economic dynamics are good for modern society, *if* the middle and working classes can balance their power. Given the decline of the working class and its absorption into the lower end of the new middle class, it has become up to this new middle class itself to balance the power of the capitalist rich. And, it will be up to the new middle class to limit the power of the new managerial upper class as well, since it is slowly and fitfully emerging as a new power class in this new society emanating from the new high-technology mode of production.

We must, therefore, ask the question again: can the new middle class assert itself as a class and act as an Aristotelian middle class mediating between the rich and the poor? Or, will the new middle class become locked into the elite-hierarchic authority structure of the modern giant bureaucratic organization?

The polis or the ancient Empire – which will be the model for the modern world?

Part II

The Polis Analogy: The Democratic Potentialities of the New Middle Class

4 The New Middle Class as an Aristotelian Base for Democracy

THE DECLINE OF THE OLD MIDDLE CLASSES

Especially in the United States, but also in Western Europe, the dramatic decline of the small business, small farmer, and independent professional classes has been well documented and is hardly controversial.[1] Family farms produce a lower percentage of total food production each year; small store owners are replaced by franchise chains in most of the new shopping centres. Supermarkets, and now, even franchise all-night grocery stores, are taking a larger and larger share of the market. This latter process is occurring more slowly in Europe than in the United States, but on the other hand, in Europe, health and service professionals are declining more quickly (as independent small businessmen), since they are becoming absorbed into the 'socialized' service sector. In the United States, of course, doctors, lawyers, accountants, and other professionals are rapidly becoming absorbed into hospital, court, corporate, and government bureaucracies.[2]

Mills was correct in predicting that the trend would continue in the direction of the decline of small business, small farmer, independent professional strata, and their incorporation, as salaried employees, into large-scale bureaucratic organizations.

It is not that small businesses cannot still be remarkably profitable. In fact, they can often be more profitable now than ever. 'Middle men' still flourish in a variety of business areas; restaurants and boutiques may be very lucrative businesses; independent doctors and lawyers still prosper. However, these strata, which make up the small-business middle class, no longer exist as the majority class in society. Mills was on target in predicting the decline of the 'old middle class'. (This class is, however, on the rise in the post-communist societies, especially in Asia, and we shall discuss the ramifications of this phenomenon.)[3]

The commercial or business middle class, of course, was never a ruling class in the Marxist sense, but rather a mediating class in the Aristotelian sense. And, in fact, this old commercial middle class alternated between the blind 'false-conscious' reaffirmation of the capitalist upper class, and an angry rebellious rejection of that capitalist upper class. In American history, the small farmers and storekeepers often turned against the 'Bankers' and the 'Robber Barons'. From Andrew Jackson to William Jennings Bryan – and even to the 1960s rejection of Nelson Rockefeller in his bid for the Republican nomination to the presidency – small business candidates fought against the power of the rich, while at other times they staunchly supported them (especially as the foreign-born, industrial working class grew as a competing majority class in society).

In Germany, the small business and professional class also oscillated between identification with the capitalist rich and rejection of them. The rejection in Germany took the form of romantic feudalism and German domination over Europe. Thus, fascist, as well as capitalist, proclivities characterized this class in Germany.[4] The commercial middle class in Germany helped bring Weimar legal democracy to Germany and then helped to destroy it. During the economic, political, and military catastrophes that plagued Germany, this class, like most others in Germany, sought for extreme political solutions.

Old middle-class values and ideas are still emerging as a political force today, but now as an embattled 'conservative' phenomenon. The 'moral majority' in the United States are viewed as a radical right-wing ultraconservative force in the land. They are no longer looked upon as the backbone of our democratic culture. Their ideas no longer fit the realities of the new society, and, therefore, they are looked upon as radically conservative, because they wish to return to the older ways.

In any case, the old middle class, or the commercial small business class, is no longer a majority class. In the Aristotelian sense, this class can no longer act as the mediating class between the rich and the poor. In the Marxian sense, the ideology carried by this class will no longer have the major impact on society that it once had. The Marxists would say that the ideology of the *petite bourgeoisie* was false consciousness in any case, for it allowed the rich to rule. But, in the Aristotelian sense, the adherence of the business middle class to an ideology of law and democracy forced the industrial capitalist rich to rule within a framework of 'government by law', and within the constraints of the electoral majority of the middle class. Thus, the commercial-industrial rich could never rule in openly despotic fashion, nor could they always gain their political goals.

THE NEW MIDDLE CLASS AND DEMOCRACY

Four critical questions present themselves in terms of the new class structure that is emerging in the modern world.

First, does the new middle class carry any democratic institutions and ideologies, or are its elective affinities linked to its bureaucratic and technocratic orientation to the world?

Second, is the new middle class locked into the authority and status structure of the bureaucratic organizations? Is class action impossible for these salaried employees of the white-collar pyramid?

Third, can democracy exist if the new potential upper stratum – the power elite of the bureaucracies – does not support it (remembering that the commercial upper class did support legal democracy, or at least a mixed polity with legal authority and democratic processes of government)?

And, fourth, can the new middle class act as a mediating class between rich and poor under the structural conditions of mass technocratic-bureaucratic society?

Let us attempt to answer each of these questions within the context of modern society as it is emerging.

THE DEMOCRATIC ELECTIVE AFFINITIES OF THE NEW MIDDLE CLASS

C. Wright Mills wrote off the new middle class as a base for democracy. He emphasized the bureaucratic and technocratic orientation of this class, and its co-optation into giant bureaucratic organizational structures.[5] On this basis, Mills predicted the probable demise of democracy. In *The Power Elite*,[6] he describes graphically a system of economy and polity that has already bypassed democracy.

The economic system, the political structure, and the class structure are changing precisely in the way that Mills (and Weber) believed they would. Therefore, one must take these thinkers seriously in terms of their predictions about the fragility of democracy under the new conditions.

However, before dismissing the possibility that democracy might survive the new social–structural conditions, let us examine the possibilities that might allow it to continue to exist, or even to emerge where it does not presently exist. The democracy we will be referring to may take on a new form, just as parliamentary democracy and polis democracy were different in form from one another. But, of course, this new form of government would have to embody the ideal-typical characteristics of democracy to a greater or lesser degree in order for us to categorize it as a form of democracy. That is, participation, power limitation, and law would have to continue to be basic. By this we mean electoral participation in leadership choice, the limitation of power of governmental leaders, and rational-legal authority as supreme over the citizenry and the leadership.[7]

First, let us look at the new middle class itself as a possible foundation for a new form of democracy. Few thinkers have found anything positive to say about this emerging set of strata, which we are here referring to as the new middle class. From Mills to Marcuse,[8] from Arendt[9] to Lasch,[10] this class, its world-view and its life-style, have been vilified. There may, however, be certain democratic orientations to the world which the new middle class carries as a class, and, these democratic orientations may be linked to the structural and ideological conditions of the new society itself.

The democratic orientations we are referring to are produced by three processes: university education, the rational world-view, and the isolated, separated private sphere.

Please keep in mind as we discuss these three processes that the new middle class is actually made up of various income, status and occupational groupings, as well as differing reference groups. We shall have to differentiate between these groups where their social actions differ. Where there is a consistency, we shall then simply refer to the new middle class as a totality.

UNIVERSITY EDUCATION AND DEMOCRACY

The new middle class is a college-educated middle class. But let us make it clear from the outset that though an educated citizenry is a prerequisite for democracy – and thinkers from Aristotle to Jefferson have agreed on that – education *per se* does not produce a democratic mentality. The well-educated Argentinian middle class, for instance, until recently could not conceive of democratic processes in government, and, university educated intellectuals from many parts of the world favour authoritarian political systems of one sort or another. The well-educated German middle class of the nineteenth and early twentieth centuries was known for its anti-democratic sentiments. Further still, if the university education which the new middle class receives becomes increasingly technical, managerial, and vocational, then the university education itself, rather than fostering a new form of democracy, will simply facilitate the locking of the individual into the new 'iron cage' of serfdom.

Let us look at both the problem of technical-managerial education and the problem of the elitist-intellectual's proclivity for authoritarian political systems.

According to Marcuse[11] and other critics of contemporary university education, the educational curriculum has moved towards purely technical and vocational training and away from the classical liberal arts and sciences.

This is certainly the case. Who can deny that 'management', marketing, accounting, computer programming and 'communications' are the most popular majors today in the US, Japan, and the EU countries? Or, that 'social service' majors, like clinical psychology, social work, and nursing are not increasingly vocational as undergraduate career tracks? Marcuse and Arendt warned that university education would no longer produce intellectually broadened, culturally oriented individuals, but rather narrowly specialized technicians (absorbed into the 'pop' culture produced by the mass media). Before accepting the situation described by the above thinkers completely, let us look at the actual educational situation as it seems to be evolving for the near future.

In the United States, at least, it was the 1960s Black and student revolts that first altered the traditional liberal arts and sciences curriculum. The students wanted to eliminate the rigid requirements, which typified the college curriculum, and create a system of open electives. They also demanded that courses 'relevant' to their life problems in the new society be added to the curriculum.

Courses were added to the curriculum – courses that focused upon contemporary social problems, such as environmental pollution, family and marital conflict, religious mysticism and ethics (and with it courses in Asian cultures and religions), human sexuality, 'personal adjustment', etc. The Blacks, of course, demanded and got courses on African-American history and culture, and in reaction to Black demands, other ethnic groups demanded the introduction into the curriculum of courses on their ethnic history and culture. Later, the women's movement had a similar impact upon the curriculum as it (the curriculum) became altered in the 1960s.

Under the assault of the student revolt, the Black revolt, and the women's revolt, the curriculum of the universities was made more elastic. In the process, however, the liberal arts and sciences did suffer as students chose less of the classical philosophy, history, language, arts and science courses.

However, the 'counterculture'[12] movement did not lead to stable careers for the 1960s students. And, though the movement made a permanent impact upon social life (good and bad in its outcome) including sexual freedom and divorce, self-actualization, and women's liberation, along with drug use and abuse, for example, it did not alter the economic and political trends it had opposed.

In the 1970s there was a massive counterrevolution against the counterculture. Within the universities this led not to the reinstitutionalization of the classical liberal arts curriculum, but instead to the establishment of business, managerial, computer, and service curricula as the central focus of the university education.

Now that this counterrevolution seems to have firmly co-opted the students into managerial, technocratic, and service career lines (leaving their private lives to the counterculture), a movement to reinstitute the classical liberal arts and sciences curriculum has gained ground. This curriculum was never dropped in Europe, and is being reemphasized in the American universities, and in developing nations.

This latter is critical. For, the students – and such students are becoming a majority group in most modern societies – will be taking a good deal of liberal arts and science courses as the core of their nonvocational education. Though computer languages and managerial skills will become basic college courses, Plato and Aristotle, Bach and Beethoven, Shakespeare and Molière, Michelangelo and Botticelli, Marx and Weber, Freud and Jung, Einstein and Heisenberg will not disappear. And, in fact, courses on Asian, Non-Western, and Third World cultures, along with the study of contemporary social problems will remain equally basic within the college curriculum.

The contemporary student – as part of the emerging new middle class – will become familiar with some great music and art, will be exposed to the logic and scepticism of the philosophers, the ethics of the theologians, the rationalism and empiricism of the scientist, and the self-awareness and sociocultural awareness engendered by the social sciences. The impact of such an education should not be minimized – though I have already warned that it does not necessarily lead to a democratic world-view.

What of the technical and managerial specialization and the grim careerism that seems to have overwhelmed the new generation of students worldwide? Could this not override the liberal arts and sciences, such that they become mere frills, while the technical-managerial training produces the basic world-orientation of the new middle class individual?

There are powerful pressures in the contemporary work world and the contemporary social sphere which are forcing the modern individual to take the 'educational' curriculum as seriously as the 'vocational' curriculum. These pressures are twofold. First, they involve the profound difficulty of personal adjustment in this new 'ego-isolated', partially anomic, 'rationalized' world. Second, they involve the creative demands of the technical, managerial, and social-service career lines themselves.

Already, the profound personal difficulties that the contemporary individual is encountering in the social sphere are forcing that individual to take quite seriously the social science, life adjustment, and philosophical–mystical–ethical courses offered by the universities.

Problems of personal relationships, personal identity, marriage and family ties, community disintegration, decline of religion, environmental decay, and so on, have pressured contemporary students and contemporary adults towards a serious exploration of non-technical studies. Not only have universities sought to present such curricula, but corporations and governments have brought in consultants to help their employees (and constituents) cope with problems of personal intimacy, authority, conflict, alcoholism, drug abuse, divorce, mid-life crises, gender conflicts, etc.

The corporations, for their part, are primarily interested in profit. But employees who function cooperatively and productively are necessary for profit. Therefore, the corporations, for their own reasons, have been pressured to help foster personal adjustment, gender interaction, self-awareness and ethical humanism. Even in the communist societies (or new post-communist societies) problems of alcoholism, juvenile crime, divorce, sexual adjustment, birth control and environmental pollution have also pressured such regimes, against their ideological proclivity, to approach such problems from a rational-scientific viewpoint.

The new middle class cannot survive in the modern world with technical and managerial orientations alone. Therefore, education geared to ecological, philosophical, natural-scientific, sociological, and personal awareness will flourish.

Let us emphasize separately the philosophic-religious dimension of 'personal adjustment' as distinct from the social-science dimension.

A crisis of mysticism and of ethics confronts the contemporary individual. For the modern individual, 'God is dead', and he or she is alone in the universe. If the anthropomorphic Judeo-Christian God is dead, his ethical commandments died with him. The modern individual is thus in search of a link to the universe and an understanding of the universe in which birth, death and life become meaningful; the modern individual is also in search of a system of ethics that can guide his or her actions as they relate to others, and, present us with at least some guidelines for what is right and wrong in the modern context of social and structural realities.

The decline of religion has left the modern individual with 'existential anxiety' and 'moral malaise'. Courses in philosophy, theology, ethics and mysticism are not taken lightly by such an individual. Institutes, such as those studying medical ethics or ethics in foreign and domestic policy are expanding, and having an impact on the educational orientation of the university and graduate schools. Ethical dilemmas are being debated in medical schools, law schools, graduate philosophy programmes, and even in technological and scientific graduate programmes. Even the mass media are being asked to scrutinize their ethical and moral standards.

Finally, the decline of Puritan asceticism in the West, that accompanied the religious decline, has produced a culture of hedonism and narcissism

– as its critics have rightly termed it.[13] But, this culture of hedonism does not lead only to 'pop' culture and pornography. It has also led to a remarkable upsurge in dance, music, and art of all kinds – 'high culture' as well as 'pop culture'. Though the sensuous aspects of dance, music and art have been emphasized and exploited, the serious side of these cultural activities is also flourishing. Modern dance and ballet are in a remarkable renaissance along with disco dancing; Mozart and Pavorati draw throngs along with the Grateful Dead and the Rolling Stones; while 'King Tut' drew as many millions to his gilded display as did Linda Lovelace (a porno queen) to hers.

NATURAL SCIENCES

The general effect of scientific teachings on the new middle class has not been all negative, all technical, all nihilistic. If it has produced a dis-enchantment with religious magic, ritual, and prayer, it has not produced a rejection of pure mysticism or encouraged a nonhumanistic form of atheism.

The warnings that the natural sciences are easily perverted and that their effect on modern society produces a narrow, specialized, technocratic, value-relative ethos should not be taken lightly. However, the humanistic trends emerging from modern science should not be overlooked either. The medical sciences, for instance, have worked hard to create cures for the diseases that still attack us. Such effects should not be taken for granted, and they do represent a humanistic tradition within the natural sciences. The whole field of health sciences, from medicine to nutrition, has pro-duced a holistic approach to personal health care which has resulted in a great gain in general health and life-span for the individual. Environmental studies programmes have been established at many universities in which scientists and students study ways in which to prevent the pollution which they have helped to create. The 'Atoms for Peace' programme, though it went astray, was well-intentioned, at least in its early years, and may create some future energy sources that will be safe.

Family medicine has emerged to help counteract the negative effects of medical specialization and hospital bureaucratization. 'Natural child-birth', or better put, prepared childbirth techniques, help the modern cou-ple come through childbirth feeling stronger, united, and ready to raise a child together. My wife, for instance, was very thankful for the 'Lamaze' training. She knew how to control her own body during the childbirth process and was very pleased to have me there encouraging her. The pain was still terrible to endure, but five minutes after each of her two births (and they were difficult, twelve-hour ordeals), she felt wonderful, ate heartily, talked cheerfully, and held the respective babies. Contrast this with my mother's experiences, where, left alone, sedated, and untrained, she screamed in pain, resented her doctors, felt miserable for days after-wards, and was, in general, traumatized by the experience – though her actual pain was no greater than my wife's, and, in fact, her births less complicated and shorter in duration.

The point is that there have been some humanistic results emanating from the modern sciences. The dual trends of amorality and humanism can be best summed up and illustrated by the increasing trend by food corpora-tions to use food additives – which eventually kill us, and, to produce more whole grain and fibre-content foods – which help us to live longer and better. There is no doubt that science produces a mixed bag of bless-ings and disasters. At this point in history one cannot tell if science will help us blow ourselves out of the universe, or colonize it. Still, the human-istic side of scientific rationality has existed since the days of the Hippocratic oath, and continues to exist – Nobel, the dynamite scientist, and his prizes, epitomize the paradox.

THE CREATIVE DEMANDS OF MODERN WORK

The industrial and administrative organizations – private and public – have discovered grudgingly that narrowly trained, poorly cultured technocrats and managers lack creativity and innovative ability. They are not good problem solvers; they do not make imaginative decisions. And, where technicians and managers lack social psychological insight, they do not deal well with people in the group-settings which characterize modern organizations.

Therefore, a liberal arts and science and social-science background may be necessary as part of the training of effective technocrats and managers (upon whom demands of creative invention, creative decision-making, and insightful personnel practice are increasingly descending). Good manage-ment, technological innovation and smooth organizational functioning are the hallmarks of the new social structure . . . when it is working well.

Corporations and governments have begun to accept the need for a broadening educational experience. They accept the need grudgingly precisely because this educational experience produces philosophical scep-ticism, humanistic morality, personal insight, social awareness, and inde-pendent creative thinking. The modern organization – private or public

– does not want ethical, creative, self-aware individuals; it wants loyal functionaries, but such functionaries may be inadequate to their tasks. Therefore, a kind of dual educational process may become institutionalized, with the technical-managerial track existing side by side with the liberal arts and sciences track as the typical educational experience of the new middle class.

The dual educational process which the modern individual will experience may produce unresolvable conflicts within the individual and society. The managerial-technical training will produce a narrow technical expertise and an acquiescence to the control and manipulation of the elite-hierarchical authority structure in the work-world, while the liberal arts and sciences education will produce the sceptical, rational, independent, humanistic thinker in the private sphere.

This conflict could lead to the dual existence posited by Weber, the Frankfurt School,[14] Arendt, and Lasch, of work-world conformity along with private self-actualization, and a subsequent withdrawal from the public citizen's role entirely. The public sphere of governance and production would be left for the 'soulless' managers and technocrats, while private hedonism would be turned to for escape.

Could the conflict be resolved in another way? Will the new middle class necessarily be passively acquiescent in its public attitude? Does self-actualization always become narcissistic and hedonistic? Will the public citizen's sphere necessarily be abandoned by the new middle class and left to the Machiavellian machinations of the power elite? Let us explore some alternative possibilities. Not as devil's advocates (or should I say angels' advocates?), but as observers of the new middle class as it now exists.

The new middle class, in its short history, has already exhibited forms of social action which do not coincide with the Weberian–Laschian thesis of serf-like co-optation and self-centred withdrawal. Can we say that the new middle class has fully abandoned the public citizen's sphere when, with a few weeks' notice and *ad hoc* organization, hundreds of thousands will appear to protest some government or corporate decision? Can we describe the new middle class as acquiescent to the power elite when they risked their lives in Poland and Czechoslovakia, Russia and China in order to establish more democracy and a better standard of living? Have not technocrats risked their careers by 'whistle-blowing' when corporate decisions threatened lives through safety overrides? Have not Russian scientists and artists risked family and career to protest the despotic nature of their regime? Is joining the Peace Corps, Vista, or the Civil Rights movement narcissistic?

Given the above observations, we wish to suggest the possibility – not probability – that the institutionalization of the liberal arts and sciences, along with the increasing percentage of the population attending the universities, could produce an individual who is rational, sceptical, humanistic and cultured, and who could, therefore, become the perfect citizen within a reconstituted democratic system (reconstituted on a technocratic base, as I shall describe).

RATIONALIZATION AND THE WORLD-VIEW OF THE NEW MIDDLE CLASS

Some interpreters of Weber suggest that the main unifying theme of his work is to be found in his concept of 'rationalization'.[15] For Weber, rationalization was the identifying 'civilizational' characteristic of the Occident (and today, of the world of modernity in general).

Rationalization was for Weber a process that altered both institutional and ideological factors. In terms of the change in ideology he refers specifically to: (1) the elimination of magic and ritual from religion, leaving pure mysticism, (2) the rise of the secular-scientific world-view, replacing the religious world-view, and (3) the decline of kadi-justice, or justice based on religious codes, and the rise of legal justice based on constitutional law. In terms of institutional changes Weber refers to: (1) the rise of industrial capitalism, or alternatively, a socialist planned economy, replacing traditional agrarian, artisan, and trade-capitalist merchant economic activity, (2) the rise of legal-rational authority and the decline of traditional authority, and (3) the rise of rational-bureaucratic administration, replacing aristocratic-nepotistic administration (and replacing the ancient bureaucracies where they still existed, as in China).[16]

Weber includes much more under the rubric 'rationalization' of thought and action – even the rationalization of music eventuating in classical symphonic orchestration – however, we shall focus on the ideological, economic and political aspects of rationalization.

Weber looked back to the origins of Western rationalization and found its roots in the ancient Jewish Prophets and the ancient Greek scientists, mathematicians, and philosophers. He then viewed the entire history of the Occident – from one perspective at least – in terms of a progressive rationalization of all aspects of social action.

Let us look at the rational world-view in terms of its effects on the political life of the modern world.

What kind of politics would emerge from a new middle-class majority imbued with a rational world-view?

The secular-scientific world-view and existential philosophy may not lead only to cynicism and hedonistic (or cultic) withdrawal from the world. They could lead to a healthy scepticism and a desire for a rational humanistic political system. In the contemporary world, emanating precisely from the new middle class, one is just as likely to find organized, active political movements demanding rational humanistic actions from government as one is to find cynical withdrawal and private hedonism.

Equal rights for minorities and women, civil-libertarian (human rights) protection for all citizens in all countries, environmental protection groups, consumer protection groups, groups demanding a humane foreign policy with arms reduction and wherein torture and genocide are eliminated (if war and revolution cannot be avoided) – anti-nuclear groups and mass demonstrations for various causes – do these not characterize the political action and ideology of the new middle class as well as low voter-turnouts and the rise of the pornography industry? Such organizations as America's Watch, Helsinki Watch, Amnesty International, the American Civil Liberties Union, and the Consumers Union are growing all over the world.

The new middle class with its technical and liberal arts and sciences education and rational orientation to the world may be the best potential foundation for democracy since the rationally oriented middle class of ancient Ionic Greece – if it can develop the power to legally limit its new leadership and the institutions through which to gain greater participation with its leadership in policy decision-making, and if it can learn to harness the new mass-media-computer technology to democratic processes.

The rational world-view may push the new middle class in one of two different and conflicting directions: either the rational world-view will propel the new middle class into a new form of rational citizenship, not unlike that of ancient Greece, but mediated by the mass media, or it will become the force which acts to absorb them into the technocratic-bureaucratic cage of modern 'serfdom'.

The rational world-view could lead the new middle class to engage in rational performance evaluation of the system at large; rational evaluation of the leadership in particular; rational evaluation of bureaucratic organizations and their top management; and a proclivity for rational law as the mode of justice, authority, and political process. However, the rational world-view could also produce the adjustment of the new middle class to administrative rationality, the needs and functions of rational administrative organizations, and to a meritocracy ideology wherein rational reward seems commensurate with attained position: a kind of 'mandarin' mentality.

The situation is like ancient Greece in terms of the decline of religious and traditional authority and the rise of the rational-scientific world-view, but unlike ancient Greece in terms of the rise of heavily articulated and rationalized administrative bureaucracies pervading every aspect of modern life and absorbing most of the citizenry in terms of their work-world and their world of services.

Notice, though, that the difference between the modern situation and that of the ancient empires is great also. For, populations of the traditional empires were not imbued with the rational world-view, or disenchanted with the world, or cynical and sceptical about their leadership, or reflective about their political system and their role in it.

THE INDEPENDENT PRIVATE SPHERE AND DEMOCRACY: UNLOCKING THE CLASSES FROM THE BUREAUCRATIC HIERARCHY

From the Weberian perspective, we get the picture of the modern world run by soulless bureaucrats while the average individual withdraws from the public world of citizenship into a private world of narcissistic hedonism and mass media mindlessness. Lasch, Marcuse, and Arendt describe precisely this situation in the contemporary world. If this is happening, therefore, one might ask why the 'private sphere' should even appear in the same subtitle with democracy. The answer is that more than narcissistic hedonism is occurring in the private sphere, and the public citizen's sphere has not been abandoned by the new middle class.

First, let us establish the fact that the private sphere is more private than it has ever been in human sociohistory. That is, the individual, or the nuclear family, or the fragment of a nuclear family (single parent) is now almost completely in a state of what Durkheim called 'egoistic isolation'.[17] Durkheim meant specifically that the individual is almost completely cut off from the extended family, the church group, and the community. Even friendship groups in the modern urban-suburban sprawl become completely unstable as modern individuals move from one area of their country to another.

The work-world of the giant technocratic-bureaucratic organization also does not provide intimate personal relationships of an enduring character. It is stable only in an 'instrumental' way, the cohesion of relationships dependent purely upon role interactions and role integrations.

Because of the isolation of the individual from primary and organizational relationships, the privatized individual can become publicly deviant, or can become politically active in any cause, any group – as long as this public political activity is completely separated from the work-world and as long as work-world conformity and loyalty are maintained, and, of course, unless this political activism specifically and directly confronts the individual's work-world organization (government or corporate).

This latter – that privatized political activity must not threaten or attack the work-world organization specifically and directly – sets the limits to the freedom of action in the private sphere. The remarkable (and worrisome) anonymity of contemporary urban–suburban non-communities protects the private sphere in most cases, such that conflicts between the family and the work-world are not very common. However, it is clear that for the private sphere to be truly free in the political sense, legal protection for the privatized activities of the individual against the government and against private organizations of any kind would have to be institutionalized. We shall spell this out later.

It is said that personal loyalties and social conscience have been given up in favour of personal selfishness. Who can deny this? One sees hundreds of isolated individuals jogging alone, roller-skating alone, biking alone, moving rhythmically to music only they can hear on their private headphones. The health clubs are full of people staying young and beautiful, the divorce rate and remarriage rate are staggering, the porno industry is flourishing (with porno cassettes the largest selling video cassettes in many nations).

It is also true, however, and we should take this as seriously as the above-mentioned description, that these same privatized, sensuous, self-centred individuals are politically concerned, politically informed, and politically active – within the constraints set by the mass technocratic-bureaucratic society. What is the nature of the political activity of the modern individual, and what are the constraints imposed upon such activity by modern society as it now exists?

The new middle class is politically active in many different ways. Individuals may join: (1) single-cause groups, like pro- or anti-abortion groups or anti-nuclear power, or environmental groups or women's groups, gay groups, etc., (2) ideologically oriented groups like the Moral Majority or Common Cause, (3) democratic watchdog groups like the ACLU or the Citizen's Union, and (4) they may join *ad hoc* groups organized to protest a particular policy or problem, such as the committee to free Soviet Jews or assist the Northern Irish or protest America's involvement in Vietnam or Russia's involvement in Afghanistan.

The new middle class – young and old – proved quite willing to join government-sponsored activist organizations as well, such as the Peace Corps and Vista and now, Clinton's Community Service Corps. One should remember that these organizations did not decline in the USA until government support was withdrawn from them in the Nixon years. In fact, it was not until after the 'Nixon repression' – the shootings at Kent State and Jackson State, the verbal attacks on the mass media, the impounding of funds earmarked for social programmes and social research, and the drastic reduction of funds to the universities and the 'think-tanks' – that the turn inward towards narcissism and hedonism occurred as a general phenomenon among the new middle class in America.

The new middle class is willing to become involved in street demonstration, such as the various marches to Washington or the anti-nuclear demonstration in New York (1984) or the democracy marches in Beijing, Warsaw, Moscow, Bonn, Berlin, Paris or London. Mass street demonstrations are a continuing phenomenon of modern society. Millions can be temporarily mobilized in support of a protest of a government policy in every country in which the new middle class is emerging.

One might insist that the street demonstrations are proof of the alienation and impotence of the modern individual in mass technocratic-bureaucratic society, for, one could conclude that if the institutionalized channels of democratic participation were working (representative parliaments and elections) then street demonstrations would not be necessary.

I heartily agree with this. But, notice, this does not lead to the conclusion that the new middle class is withdrawn and selfish and anti-humanistic. On the contrary, it leads to the conclusion that the new middle class is involved, concerned, and willing to be active in the public sphere as public-minded citizens, but that the democratic institutions of state are inadequate to the demands of the new mass technocratic-bureaucratic society which has emerged.

Given this conclusion, the direction of one's scientific inquiry should be towards the attempt to formulate ways in which the democratic processes of government could be adapted and expanded in light of the new technological, organizational, and residential structure of society and in order to facilitate and incorporate the public-spirited desire of the new middle class to act in the citizen's role.

A further 'proof' of the citizen's potential of the new middle class is the fact of increasing participation within the institutionalized structure of the political parties. It was just a few years ago that the party regulars and party bosses of the Democratic Party in America bemoaned the takeover of the party by amateurs and young people. The 'democratization' of the party delegate rules and the remarkable electoral activism exhibited during primaries must be seen as a fact of political citizenship emanating from the new middle class. Such activity has occurred with 'the

Greens' in Germany, and the Social Democrats in England, the Liberals in Canada, etc. And, of course, now in Eastern Europe party participation is beginning.

The voting percentage of the populace is low in the USA (though it is high in Europe), but the level of electoral participation is very high. Within the Republican Party the Goldwater–Reagan grass-roots phenomenon and the 'moral majority' reaction to the social changes of the 1960s produced the same kind of 'democratization' within the Republican Party as that which had occurred within the Democratic Party. This latter process was at first generated by those espousing the views of the old middle class in America. But today, the conservative Republican activism emanates from the Sun Belt and the upper-middle-class suburbs of the nation in general. A similar moral majority conservative party movement has occurred in France as well.

Thus, as we have tried to establish, the new middle class has hardly 'withdrawn' into its privatized sphere of isolated existence. In fact, because of the anonymity now accorded the private sphere, and its complete separation from the work-world, this private sphere has become the sphere in which political action – according to one's personal choice – has been stimulated. For every jogger, there is a marcher, for every cocaine sniffer, a cause worker, and for every 'VCR orgy', a demonstration. Narcissism and humanism do not negate each other in the ideology and social activity of the new middle class. In fact, one should remember (and Arendt does not mention this)[18] that narcissism and public citizen activism went hand in hand in ancient Greece as well. Aristotle rails against hedonism[19] and Plato dreams of the re-introduction of Spartan asceticism into Athens,[20] but democracy and high culture flourished along with hedonism and cultism in Ionic Greece. No one was more vain or self-centred than an Ionic Greek orator or athlete, and no one was more committed to democracy.[21]

What of the constraints imposed by contemporary mass technocratic-bureaucratic society on the political involvement of the new middle class?

THE CONSTRAINTS ON THE INDEPENDENT POLITICAL ACTION OF THE NEW MIDDLE CLASS

The basis of much of the pessimism derived from the Weberian and neo-Weberian prediction of the future revolves around the non-independent nature of the new middle class (as opposed to the old middle class of independent shopkeepers, farmers, small businessmen and professionals). The new middle class is described as locked into the hierarchies of the bureaucratic organizations, as salary-dependent, and subordinated to the power of the elite managers at the top.[22]

This description is true to the facts. The new middle class cannot be politically free as long as it is constrained in its speech and participation by loyalty-inhibitions emanating from the work-world. The modern organization – private or governmental – does not care what you do in the private sphere as long as you are not disloyal to it (the organization). One cannot cause trouble for or inhibit the goals or profits of one's organization or government bureau. Public opposition to one's organizational policies brings down swift punishment against the individual. Not only loss of job, but loss of career may result because of permanent blackballing. (In communist countries, of course, mental institutions and jails had been the punishments for such activities.)

What is needed to protect the individual in this regard is nothing short of a new Bill of Rights relating to citizens' protection for those individuals working within public or private organizations,[23] and the establishment of a federal Ombudsman system[24] to enforce the new bill of rights and protect the new middle-class citizens from the power elite and hierarchy of modern giant bureaucratic organizations. We shall discuss these newly necessitated democratic institutions in detail later on.

It may be that protecting the individual from the large-scale technocratic-bureaucratic organization is not enough. The actual democratization of these organizations may have to occur.[25] It may not be practical, or even possible to do this in terms of the organizational efficiency of modern bureaucratic administration, but it may be necessary to try.

THE MASS SOCIETY

The word 'alienation' has acquired a new meaning in the modern context. Along with its formal Marxist definition, it has now come to represent the feeling of political impotence that the modern individual acquires because of the vastness of the society at large. The individual feels so distant from the centres of power that it seems that nothing he or she could do would influence the decision-making process.

The modern individual in mass society feels far away from the Power Elite – the corporate executives, politicians, generals and others – who make the decisions. And the individual feels distant from his or her elected representatives as well. One of the reasons contemporary individuals join street demonstrations is that they do not feel their local representative either: (1) knows what they desire, or (2) could do anything about it if they did.

There are two levels of alienation, then, occurring. One in regard to the Power Elite, the other in regard to the elected representatives.

Two different solutions are required in this regard. One solution would involve the limitation of power and tenure, and the possibility for recall of, or election of, the Power Elite. The other involves utilization of new communications technology in order to link citizens more closely to their representatives. The possibility for cable-TV town meetings, the growing popularity of interactive radio and TV talk-back shows and issue-oriented referenda will be analysed in this regard, along with schemas for democratization of bureaucratic organizations.

THE MASS MEDIA

In terms of the mass media, both the content and the process of political action have been significantly altered. Free speech and debate along with contact between citizens and candidates for office have been literally mediated by television, radio, news magazines and newspapers – one feels as if the experience is immediate and personal, but of course it is actually distant, and can be manipulated.

This situation varies from country to country in terms of the parameters of political debate that reach the public through the media, but that the public receives its political information from the media, as well as its perception of the candidates, is a fact in all modern countries.

As MacLuhan has suggested[26] the process of politics has been altered by the structure of the media. Television, for instance, alters the kind of oratorical style a political leader must use in order to be effective. Low-key oratory and quick, packaged 'sound-bites' rather than detailed policy statements emerge from this medium as effective communication devices. Television makes leaders appear larger than life, yet ordinary, immediate, thought they are a universe away. The good and bad habits of candidates are viewed as if through a magnifying glass, causing us to accept or reject the candidate in a very unrealistic assessment of their abilities. The viewer becomes the passive recipient of fantasies and carefully packaged perceptions from an active political world beyond his or her reach. This is hardly a citizen's civic connection to the political process.

As Mills suggests, leaders become 'stars' or 'celebrities' – and celebrities become leaders – not charismatic figures for whom one would fight and die for, but fantasy figures existing in a media-manufactured world unconnected to one's ego-rational control, generating excitement and awe without generating comfort, confidence or morality.

In terms of content, whichever groups have access to the media dominate it with their viewpoint, overwhelming the citizen with their 'spin' on a given policy issue or candidate choice. Obviously, some system of fair access to the media must be guaranteed to all political aspirants. Political policies must be stated in their full spectrum of opinion. Political debates and 'commercials' must be regulated legally. Either free access to television or subsidized time must be granted equally to all legitimate candidates.

In terms of process, the use of more sophisticated media technology may be the answer. Communications technology now exists which allows the individual to participate in, and become active with, the media. This could help to remove the passive fantasizing of the viewer and media-charismatization of its celebrities. Cable television town meetings may become the democratic institution of the future high-tech world.

To sum up this section: the new middle class, with its university education, rational world-view, and anonymous private sphere can become the class base for a new democratic citizenry if their civil liberties are protected in the work-world, and if new institutionalized channels of democratic participation, limitation and law are created and adapted to the new mass technocratic-bureaucratic world. We shall discuss these potentialities in greater detail in the last section of this volume.

5 The New Middle Class and Law

INTRODUCTION

Having discussed the potentiality for the new middle class to participate as a responsible, rational, public-minded citizenry, let us now turn to the problem of law. Since law was linked in its origins and its entire history with the commercial classes, since it was carried by these classes alone, will law decline and be replaced by bureaucratic rules and regulations? Or will law and legal authority be supported and sustained by the bureaucratically based new middle classes?

We must understand that in the ancient world, where bureaucratic administration was dominant, the 'law' of the merchants was pushed aside and failed to evolve towards constitutional law and civic law. Instead, a rational-style of 'mandarin' administration, backed by the authority of the divine kingship, emerged as the typical form of authority. Once established, this form of traditional authority prevented legal authority from emerging.

In the modern world, as bureaucratic administration broadens and deepens within the state and the transnational organizations, will legal authority be eroded and eventually replaced?

THE CONFLICT BETWEEN LEGAL RATIONALITY AND BUREAUCRATIC RATIONALITY IN THE MODERN WORLD

We have established that the new middle class carries with it a rational world-view. Since law and legal authority are rational, one might think that a simple one-to-one compatibility might arise between the rational world-view of the new middle class and legal-rational authority. Unfortunately, it is not that simple. For, as we shall show, bureaucratic organization possesses its own form of rationality. And, the rationalities of the two are not only different, but antithetical.

It must be pointed out at the outset, however, that bureaucratic organization carries with it an irrational component as well as a rational component. The rational component is related to the efficient administration of complex tasks (often involving a mass population). The irrational component has to do with the power, tenure, and succession to elite leadership, and, in a different vein, with the rigidity of bureaucratic rules and regulations and their unadaptability to unique individual cases.

We have included 'legal authority' as one of the bases for democracy. From Aristotle to Montesquieu, the rule of law over men was considered as one of the prerequisites for a stable democratic form of government in which the protection of the individual and the limitation of the power of the leadership receives equal weight with the participation of the people in the political process.

Legal democracy, as opposed to pure democracy (decision-making by the body of citizens without constitutional–legal constraints), is the form of democracy upon which capitalist industrial democracy is based. Of course, the direct democracy of the Greek city-state has given way to the representative democracy of the nation-state.

In the mass nation-state of modern industrial society, the rule of law – in terms of the protection of the citizens and the restraint of the leadership – is all the more important since the seat of government is more distant from the populace and less directly connected to their immediate control. Therefore, the decline of law could signal the decline of democracy in mass industrial society.

Why should 'legal authority' decline in modern society? The answer to this question is found in the rise of bureaucratic organization in every public sphere of modern life.

It may be that the conflict between legal and bureaucratic authority constitutes the central problem of the twentieth century.[1] Legal authority could eventually be overwhelmed by bureaucratic necessity in the modern technocratic state. The need for complex coordinative administration in both the political and the economic spheres is so compelling that it may overcome all resistance to it. And since the *laissez-faire*, parliamentary form of government cannot in itself provide technocratic administration, the bureaucratic state may eventually take its place. Let us look at why this would be problematical.

LEGAL RATIONALITY

Law vs Administrative Rules and Regulations

The political result of legal authority is to fully establish as central a set of rational processes of legitimation. Under a system of legal authority, government by law replaces familial and traditional procedures of government; clan justice; Kadi-justice (religious courts); the justice of the kingly court; and bureaucratic or administrative justice.[2]

The most significant aspect of secular-civil law, even where constitutionalism is surrounded by ideology, is that it is open to change and subject to rational debate. Amending and abolishing procedures are institutionalized and are constantly encouraged. Furthermore, at least in principle, the breaking of a law by an individual also sets in motion a process of rational debate and rational evaluation for the determination of guilt, and if guilt is established, for the determination of punishment.

While it is clearly understood that no process of amending or abolishing norms or world religious codes existed in traditional societies, and that kingly decrees were hardly open to rational debate, it is not so clearly understood in the modern world that bureaucratic rules of administrative procedure are different from laws. Such rules are not open to rational debate; they are created by elite cliques in relatively closed sessions; no institutionalized channels for their amendment or revocation exist, and their rationality is not that of the entire political entity, but only of the organization whose hierarchy they support.

This clash between legal procedures (laws) and bureaucratic procedures (regulations or red tape) is central to the problem. It is extremely difficult to extend the rule of law over bureaucratic organizations because the wall of organizational rules often prevents the individuals within bureaucracies from responding to law at all. The very efficiency of the bureaucracy is related to its rational rules of procedure. 'External' laws may be seen as an impediment to efficiency and a threat to the hierarchical authority structure – and they are therefore resisted.

Political Participation

Under legal authority, rational debate and discussion are institutionalized and encouraged as the proper process of decision-making. Of course, whether this is a direct or representative process is critical. But, even where this is a representative process in a 'mass democracy', rational debate at the local level and numerous local and national elections do produce continuing discussion within the population at large, as well as among the representatives themselves.

Whether such debate is actually directed and limited by a corporate-controlled media which establishes the focus and boundaries of debate, is crucial; this is also truer than American liberal theorists have been willing to admit, for alternative media of expression do exist, but they have very 'small voices'. Nonetheless, in nations where legal rationality exists, the idea of debate as open to, or even incumbent upon all, also exists. Where legal rationality does not exist, the controversy over whether debate is free or controlled never arises, for debate is repressed directly.

In the same vein, whether the representative process in mass democracies actually produces popular participation in political decisions has been much debated, not only currently, but even in the days of Hume[3] and Rousseau.[4] Whether mass representative democracy is truly participatory or whether it is over-influenced by the rich to produce a 'veiled oligarchy', as Marx asserted, is obviously a critical issue as well. In the present context, however, it is enough to note that where legal rationality exists, direct and representative participation are held up as ideals to be maintained at all costs.

In its decision-making process, on the other hand, bureaucratic rationality prohibits open discussion and debate as being administratively inefficient and disruptive to the long-term planning of the bureaucratic organization. It is interesting to note in this regard that even Solzhenitsyn, who fought so hard for free speech and other freedoms in the Soviet Union, has criticized lawful parliamentary procedure as being too chaotic and inefficient (1981). Apparently, life in a huge bureaucracy effects even those so oppressed by it as to risk their lives for civil liberties. That reactions like those of Solzhenitsyn may not be atypical is, of course, one of our major concerns.

The Limitation of Power and Tenure of Leaders

Under a system of legal authority, limitations of the power and tenure of political leaders are institutionalized. The recall of leaders who are poorly evaluated, or who break the law, is also institutionalized, through electoral or court procedures. Furthermore, either 'amateurs' or lawyers are preferred as office holders.

Conversely, few limitations are placed on bureaucratic administrators. Life tenure often replaces limited tenure, and, professionally trained specialists are logically preferred.

Rational Succession Procedures

Since tenure of office is limited, the procedures for leadership succession occur constantly and therefore has to be rationalized. Though powerful groups and classes have often sought to appropriate both access to office and the electoral procedure itself, thus far they have always failed to overcome the ideology of legal authority.

Naturally, attempts to control this rational electoral procedure and to influence the incumbents continue, not only by powerful groups, but also by any organized groups in society. Again, whether certain power-groups (i.e. the rich) succeed in over-influencing the electoral process and the incumbents of office is a critical question for debate between Marxists and liberals. By contrast, electoral succession procedures are missing from bureaucracies. Leadership and succession are determined through a combination of expertise and purely 'Byzantine' power and manipulation techniques.

Access to Leadership Offices

Within a system of legal authority access to all political leadership roles may be open to all citizens or legally limited to a specific class of citizens. Technically, legally closed leadership offices can be established. For instance, property and educational qualifications have existed.

In bureaucracies, the situation as to access to leadership offices is quite different from that of legal authority, though they do exhibit the similarity of open access leadership roles.

In bureaucracies, however, open access to leadership offices is not linked to the natural law ideology of the equality before the law of all individuals and the duty of amateur citizens to participate in the decision-making processes of government and in the leadership offices of state. In bureaucracies, it is specialization and technical expertise which lead to the system of open-access leadership offices. Nepotism and aristocracy are eliminated because familial and aristocratic ties do not guarantee the kind of expertise demanded by the bureaucracies for their proper administrative and technocratic functioning. Thus, though open access leadership roles do exist, they exist within the logic of bureaucratic rationality, rather than legal rationality.

Furthermore, and perhaps even more importantly, the selection process in bureaucracies is directed by those at the top of the organization. Thus, the democratization of bureaucracies is rarely fully extended, since leadership cliques entrenched at the top can and do direct and limit the access to leadership positions.

On the other hand, the elimination of nepotism is democratizing in its effect. It must be mentioned again, however, that this is accomplished, not out of the legal-rational principle of open leadership succession, but because of the technical needs of specialization. A gulf of difference separates the motives.

Legal Authority and the Separation of the State from the Economy

Under a system of legal authority, economic action may be relatively free of state control. This may be a logical rather than historical connection within legal rationality, for the separation of economic from political activity does provide an important limitation on political power. Even if the Marxists are correct in arguing that the rich capitalists use this separation to control the polity, because this control is indirect and must be 'veiled' (where actual oligarchy is illegal), both the influence of the middle and lower classes, and the civil liberties of all, are maintained.

On the other hand, the bureaucratization of the economy leads to administrative centralization and the linking of economy and polity. The power of political leaders could be greatly enhanced by this, since the power of the state – that is, military and police power – is now joined to the power of the economy – that is, the control of the necessities of life-sustenance.

BUREAUCRATIC RATIONALITY

Bureaucratic rationality is linked to and evoked by administrative and large-scale organizational efficiency and stability over long periods of time.

Specialized Experts vs Amateur Officials and Lawyers

Whereas legal rationality encourages amateurism and mass involvement, with the only essential specialists being lawyers themselves, bureaucratic rationality demands the installation of specialized experts at every level of its organization structure.

Hierarchy vs Collegiality

Most critically, a hierarchical form of authority, with power emanating from the top down and with strong clique-authoritarian tendencies at the top, replaces the 'collegial'-democratic, limited power structure of legal authority.

The 'Power Elite' vs Citizens' Participation

Decision-making is not a process open to debate, persuasion, or even mass manipulation, but rather is a process shrouded in secrecy, emanating from the top, or the elite, of the hierarchy; the decisions are not debatable and are closed to the majority below. Suggestions may be solicited from the lower ranks and expected from the middle levels of the hierarchy. Such input is in fact necessary if those at the top of the hierarchy are to make efficient long-term decisions. This input, however, can be accepted or rejected at the will of those at the top of the hierarchy.

The Power Elite vs The Limitation of Power

No institutionalized limitation on leadership powers exists, nor could limits be forced upon leaders by those below. No institutionalized channels of limitation or recall exist in bureaucratic-hierarchical authority structures. Leaders are not accountable to, or removable by, those at lower levels, but are checked only by others in the power clique at the top – power against power, manoeuvre against manoeuvre, manipulation against manipulation.

Succession Through Power Struggle vs Rational Electoral Procedures

One of the great anomalies of bureaucratic rationality is that rational succession rules do not exist for the top leadership positions in a bureaucracy. Clear and rational criteria for promotion and hiring exist at all levels of bureaucratic organizations except at the top. Of course, throughout the hierarchy, promotion is based upon 'politicking' which, though stripped of direct coercive power in most cases, is nonetheless psycho-socially ruthless. In contrast, with the electoral judgement of the populace placed between the combatants for leadership positions, as in legal rationality (even though the people can be manipulated by demagogues and media), just the fact that leaders must take their struggle to the people and must find grounds on which to manipulate them lessens the totality of the Machiavellian impulsion. And, venal as the senators of Rome may have appeared, the emperors, after Octavian, who did not have to go to the people for affirmation, made them look like angels of grace.

Thus, within bureaucracies, qualifications for high-level succession are based less on specialization and more on byzantine politicking. Some minimal specialized expertise is demanded of those at the top. But beyond this minimum necessary level of administrative expertise, succession to high bureaucratic posts is determined purely by power plays and manoeuvring. The lack of rational succession procedures is, of course, one of the irrational elements of bureaucracy. It may, in fact, be the greatest flaw in bureaucratic organization. It is also a regression away from the rational succession procedures of legal authority.

If the bureaucracy is linked to the political state and top leaders have the coercive power of the state at their disposal, then, of course, the byzantine manoeuvring becomes not merely a power play where losers drop out, but a coercive struggle where losers are tortured and murdered along with their families and followers. Machiavellian politics replaces the 'executive suite' when bureaucracy and the state are synonymous. In neither case, of course, do legal-rational succession procedures operate.

Life Tenure vs Limited Tenure

The whole Enlightenment and Polis tradition is negated by bureaucratic life tenure. Accountability and limitations are removed and a whole government of 'little dictators' is created.[5]

The Repression of Charisma vs the Institutionalization of Electoral Charisma

Charismatic leadership is inhibited, rejected and expelled under bureaucratic authority.[6] Under legal authority charismatic leadership is not necessary. Though legal authority is made to operate without the need for charismatic leadership, its own institutionalized procedures do allow for the expression of charisma in its safest and most controllable form – elections. Weber saw 'charismatic plebiscitarian democracy' as a possible saviour of the modern legal-bureaucratic state. Unfortunately, there are problems within this 'solution' which Weber did not foresee,[7] for there is a fine line between the demagogue and the dictator. Nevertheless, with legal authority in place, demagogue-electoral charisma can be safely utilized for the good of society.

However, bureaucratic rationality absolutely opposes charismatic leadership because charismatic leaders by their very definition tend to break people loose from the regulations, institutions, and ideologies of the organizational and societal structures which they operate within. This is too threatening for bureaucracies to handle. They prefer stability and longevity to dynamic and dramatic success. The bureaucratic army prefers the Eisenhowers to the Pattons, the bureaucratic corporation, the MacNamaras to the Iacocas.

Within legal rationality, though, no president is above the law. However, the populace and, in some cases, their representatives, cherish the charismatic leaders, such as Washington and Roosevelt, or Disraeli and Churchill, while denigrating mundane leaders such as Polk or Coolidge, Gaitskill or Heath. Still, these plebescitarian leaders are contrained within constitutional law.

Organizational Rules vs Laws

In bureaucracies, it is the organization itself which makes the rules. The people and their representatives make the laws under a system of legal authority. Of course, organized political 'interest groups' linked to class, ethnic, reference and other power-strata intervene between the people and their representatives in the lawmaking process.[8] But laws are open to public debate and discussion; they can be changed, newly initiated, amended and continually scrutinized. On the other hand, bureaucratic rules are made in secrecy, are not subject to debate, and are not open to alteration (except by the power elite of the bureaucratic organization itself).

Systems of appeal, or grievance procedures, do exist within some bureaucracies, but the judges and juries of such appeals are members of the organization itself, and, therefore, a neutral, rational judgement is difficult to achieve.

Of course, bureaucratic regulations can be appealed to in the legal rationality of the court system. And, in such case, the rationality of law clashes head-on with the rationality of administration. However – and this highlights the conflict between these two types of rationality – where this has occurred, bureaucratic organizations have exhibited a remarkable disregard for the decisions of the law courts where such decisions go against them and where such decisions demand the alteration of their rules or procedures.[9] The rationality of the law courts and the protection of citizens' rights is usually viewed by bureaucratic organizations as a threat to their efficiency and stability. Whether such bureaucracies are public or private matters little in terms of their antagonism to legal authority.

Within the governmental structure itself, the tendency of the bureaucracy to resist electoral leadership and inhibit the executive task of carrying out enacted laws has been increasing over the years.

To sum up. Though bureaucracy is procedurally and organizationally rational, it operates politically in a highly irrational manner. It discourages participation, lives on secrecy, prevents the limitation of power, inhibits checks and recall of its leaders, creates hierarchy, and destroys individualism and collegiality. It tends to produce not consent, but acquiescence.[10]

The establishment of bureaucratic authority tends to undermine legal authority precisely because it inhibits the rational processes of legitimation which legal authority makes central.

Legal authority establishes the rational procedures for consent-getting – i.e. participation, limitation and law – and allows for the expression of charismatic leadership (containing the irrational bond of the charismatic leader to the populace within the restrictions of constitutional law).[11]

Bureaucratic authority leans on the irrational processes of consent-getting – such as institutional charisma, office charisma, and manufactured charisma, co-optation and organizational rules and regulations.[12] Bureaucracies sometimes lean on forms of coercion as well where these other processes fail to produce acquiescence and consent. Blackballing, blacklisting – even physical and psychological violence – may be resorted to by bureaucratic leaders against individuals who challenge their organization.

It is possible, given this latter analysis, that the ongoing bureaucratization of the economy, polity and social service sphere could undermine legal authority entirely. It is also possible, however, that in the rational world produced by high-technology industrialism, legal authority – because of its rational character – may be retained.

Let us now look at the possibility that law may be retained in the modern world, and that the new middle class may become an active carrier of legal authority.

THE POTENTIAL FOR LEGAL AUTHORITY TO TRANSCEND ITS COMMERCIAL–CAPITALIST ORIGINS

Criminal Justice and Rational Law

It may be that in modern society law is still necessary as one kind of rule system for the maintenance of group order. For it has become clear that bureaucratic rules and regulations, which function effectively within the sphere of administration, will not function properly within the sphere of judicial action or within the sphere of leadership succession and political control. In the first case, it seems clear by now that where large-scale bureaucratic organizations – whether they be corporations or government bureaus or communist party-directed bureaucracies – accuse an individual of rule-breaking and put that person on trial within the hierarchy of that bureaucratic organization itself, the individual and the peers of that individual do not accept the judgment of guilt or the punishment meted out as either 'just' or rationally determined.

Such a judicial system is clearly viewed as controlled by the accusing agency, and therefore, loaded against the individual whose case is in question. For, the accuser, the judge, and the jury are all one. The power of the organization is simply brought to bear on any individual accused of breaking the rules.

An educated, rationally oriented populace can hardly accept such a judicial system.

When we come to the judicial procedures engaged in by bureaucratic organizations, we find that they are not only resented, but rejected outright by large segments of the modern citizenry. The outrage expressed by those 'tried' in communist countries such as Russia and China is already known worldwide. But the outrage expressed by those 'tried' by their corporation, government bureau, school, hospital, or mental institution in the West is only beginning to be known. Space does not allow me to provide cases of the latter situation, but I refer the reader to the literature on bureaucratic and corporate harassment against 'whistle-blowers' and other such 'troublemakers'.[13]

It may be that Law – rational, debatable, amendable – as the most rational form of judiciary, will become not only necessary, but consistent with the rational character of the new middle class, and therefore 'carried' by such classes (if they are not fully coopted into, and ideologically manipulated by the managerial elite).

Civic Regulation and Rational Law

Where laws guide civic action, most modern citizens can evaluate such rules in a rational manner. If they judge the laws to be good, they obey them. If they judge some laws as bad or ineffective, they know that they can change those laws through rational procedures. The citizen can protest a law, defy a law, or attempt to amend a law, and this rational process of rule establishment and rule alteration is encouraged in legal-democratic societies – the procedures of which are explicit – conservative enough to prevent anarchy, yet liberal enough to allow for change.

What would replace law and legal authority as the rule system for the maintenance of order in the modern world? By logical extension, bureaucratic rules and regulations should replace law. And, the authority of the bureaucratic hierarchy itself would replace legal authority. Logical extensions, however, rarely occur in human history. For mathematical logic makes its predictions from a static set of premises, while human interactions always produce new social phenomena which alter the equation.

The point is this: while civic laws appear as rational and amendable to modern citizens, bureaucratic rules and regulations appear as irrational – both to those serviced by the hierarchy and by those in the hierarchy implementing such rules. In most cases, neither the client nor the line employee has the slightest idea what the rationale behind the bureaucratic 'red tape' might be. The client and the employee follow such rules and regulations only because the service requested necessitates it, or their job depends on it.

Further, unlike laws, the individual confronting bureaucratic rules and regulations often finds them to be arbitrary and unfair when applied to his or her particular case. Therefore, modern populations do not perceive bureaucratic rules and regulations as rational, although they do recognize their integral connection to the stability, efficiency, and longevity of the giant organizations.

For example, few individuals at an American motor vehicle bureau find the forms and structure of the rules and regulations to be helpful for them. While realizing that millions of individuals are served by the motor vehicle department, most citizens dread going, and believe they will be unfairly treated in some way.

The same can be said for the European government offices, which caused so much discomfort amongst post-World War II citizens that the Swedish institution of Ombudsman was institutionalized throughout Europe. (We shall discuss the Ombudsman later on.)

Here, let us establish that populations with a rational world-view perceive the elite–hierarchical structure of power and authority as irrational in two ways: one, through performance evaluation, where the performance fails to live up to rational expectations and standards, and two, when the power of the bureaucracy impinges upon them and they cannot defend their rights or protect their interests.

Habermas, for instance, in his book *Legitimation Crisis*,[14] points out that the modern citizen will engage in rational performance evaluation of government and its services. According to Habermas, the citizen will, inevitably, be critical of government and rejecting of its service record. Thus, because the 'performance' of government will be judged negatively by the majority of the population, a withdrawal of 'consent' will occur, creating a legitimation crisis.

Going in the opposite direction from Habermas, one could suggest that the rational performance evaluation of government and corporate bureaucracies will give the citizen a healthy distance from the acceptance of bureaucratic authority. In this sense, the sceptical and critical performance evaluation does not create a 'legitimation crisis' for legal-representative democracy, but rather for government, corporate, and service bureaucracies.

Since the average citizen fears bureaucracies and understands their power-potential, the protective mantle of the law becomes one of the institutions which the new middle-class citizen will continue to deeply appreciate.

ARISTOTLE ON THE RATIONALITY OF LAW

As discussed earlier, it was Aristotle who first stressed that government by law was superior to government by men. Aristotle specifically linked law with the rational portion of our human behaviour and warned against the irrational side.

> He who commands the law should rule may thus be regarded as commanding that God and reason alone should rule; he who commands that a man should rule adds the character of the beast.[15]

Aristotle also characterized law as rationally amendable and, therefore, rationally adaptable.

> Law does the best it can: it trains the holders of office expressly in its own spirit, and then sets them to decide and settle those residuary issues which it cannot regulate, 'as justly as in them lies.' It allows them to introduce any improvements which may seem to them, as the result of experience, to be better than the existing laws.[16]

Aristotle warned that even in a democracy, if the rule of law is not sovereign over the will of men, democracy breaks down into mob rule, wherein the majority acts like a tyrant, or, worse yet, it breaks down into tyranny itself, wherein a demagogue gains sway over the majority of the citizens.

> A fifth variety of democracy is like the fourth on admitting to office every person who has the status of citizen; but here the people, and not the law, is the final sovereign. . . . Demagogues arise in states where the law is not sovereign. The people then become an autocrat made up of many members. . . . Law should be sovereign on every issue, and the magistrates and the citizen body should only decide about details.[17]

In Aristotle's conception of human nature, the rational and the irrational sides are recognized. The law serves to enhance the rational behavior of humans and inhibit the emotional and irrational portions.

Bureaucratic rules and regulations, on the other hand, do not encourage a rational evaluation and amendation process. Bureaucratic hierarchies do not allow for the easy discussion and debate over their rules. The rules are made by the elite managers in secret, and they are to be carried out, by employees and clients, without debate.

'Mandarin' rule eventuates in a 'byzantine' maze of rules and regulations enforced by a distant elite, whereas constitutional laws, being debatable and amendable, engender a rational approach to civic rules and regulations and to civic leadership.

But, what of the problem of 'order'?[18] It is often said that democracies are disorderly, and that dictatorships and bureaucratic rule produce a more lasting and stable social order.

THE HOBBESIAN PARADOX, AND LOCKE'S SOLUTION TO IT

Hobbes, of course, was concerned with the problem of order. Writing as he was during the period of turmoil and near-anarchy of the English Revolution (the Civil War),[19] his main concern was stemming the disorder that had shattered English political life.[20]

As is well known, Hobbes favoured a strong, unified sovereign government, which he believed could bring order back to England. Hobbes was annoyed by the Aristotelianism that was popular in England at that time – especially the portion of Aristotle's work which favoured a 'mixed polity',[21] for Hobbes believed this would allow for too much disorder. According to Hobbes, Aristotle was 'the worst teacher that ever was'.[22]

Hobbes believed that a unified sovereign would eliminate the potential for civil war – that a powerful unified sovereign was necessary because the irrational aspects of human nature dominated the rational portion.

> for the Passions of men are commonly more potent than their Reason. . . . It is true, that certain living creatures, as Bees and Ants, live socially one with another (which are therefore by Aristotle membered amongst Political creatures;) and therefore some men may perhaps desire to know, why mankind cannot do the same. To which I answer.
>
> First, that men are continually in competition for Honour and Dignity, which these creatures are not and consequently amongst men there ariseth in that ground, Envy and Hatred, and finally warre . . .[23]

According to Hobbes then, humans, unlike ants or bees, even though they possess reason, will act emotionally and irrationally towards one another. In fact, reason only adds to the potential 'war of all against all'.

> amongst men, there are many, that think themselves wiser . . . better than the rest; and these strive to reform and innovate . . . and thereby bring—Distraction and Civil Warre.[24]

Therefore, only a powerful unified government can bring order to human society.

> The agreement of these creatures (ants and bees) is natural; that of men, is by covenant only, which is Artificial; and therefore it is no wonder if there be somewhat else required (beside covenant) to make their agreement constant and lasting; which is a Common Power, to keep them in awe. . . . The only way to erect such a Common Power . . . is to conferre all their common power and strength upon one Man, or upon one Assembly of Men. . . .[25]

138 *The New Middle Class and Democracy*

Notice the paradox here. From Hobbes' line of argument one would have expected the theoretical grounding for absolute monarchy, or even dictatorship (tyranny). But this is not the case. Hobbes inadvertently establishes the groundwork for the legitimation of a legal-democratic polity. For, Hobbes, imbued with the spirit of the Greek and Roman theorists whom he criticized, and with the spirit of his own class and his own epoch, backs away from his own characterization of human nature (in a state of nature) in two key aspects:

First, the unified, powerful sovereign could be the parliament ('one Assembly of Men'), just as well as the King ('one Man'), but no mixed polity, either in the Aristotelian or Polybian sense.[26]

Second, the individual retains the right of self-protection from the sovereign, and if threatened by the government with death or interment, may defend himself or flee to maintain his life.

we are to consider, what Rights we passe away, when we make a commonwealth; or . . . what liberty we deny ourselves by owning all the Actions (without exception) of the Man, or assembly we make our Sovereign. For in the act of our submission, consistent both our Obligation, and our Liberty.[27]

The End of the Institution of Sovereignity (is); namely, the Peace of the Subjects within themselves, and the Defense against a common enemy. [But] Covenants, not to defend a man's own body, are voyd. Therefore, if the Sovereign command a man (though justly condemned), to kill, wound, or mayme himself; or not resist those that assault him; or to abstain from the use of food, ayre, medicine, or any other thing, without which he cannot live, yet hath that man the Liberty to disobey.[28]

This idea – the idea of the citizen's self-defence against governmental power – is at the heart of the rational component of natural law philosophy (which Hobbes criticizes in the earlier chapters of *Leviathan*). This right of self-defence, as we shall see, remains in the modern context the basis for the maintenance of law over against the power of the modern state. Before elaborating on this, let us look at still another Hobbesian paradox.

The second paradox within the Hobbesian logic was pointed up by Locke. That is, if humans, in a state of nature, act violently, bestially, and jealously, and are thwarted in such action only by a strong government, who thwarts the leaders within the government from acting in such a way? In a strong unified sovereign, there are few constraints on the actions of the human beings who make up the government. Therefore, they, the governmental leaders, could and would act violently, bestially and jealously.

Locke postulated, therefore, that just as humans need external constraint to prevent them from committing violence against one another (i.e. government), so too the humans within government need external constraint to prevent them from acting violently, bestially and jealously against the citizenry (i.e. law).[29]

> he who attempts to get another man into his absolute power does thereby put himself into a state of war with him. . . . For I have reason to conclude that he who would get me into his power without my consent, would use me as he pleased . . . and destroy me too, when he had a fancy to do it . . . (or) make me a slave.[30]

> He that thinks absolute power purifies men's blood, and corrects the baseness of human nature, need read only the history of this or any other age, to be convince of the contrary.[31]

> mankind will be in a far worse condition than in the state of nature if they shall have armed one, or a few men, with the joint power of a multitude to force them to obey at pleasure the exorbitant and unlimited decrees . . . without having any measures set down which may guide and justify their actions. . . .

> Those who . . . have a common established law and judicature to appeal to . . . are in civil society one with another; but those who have no such appeal – I mean on earth – are still in the *state of nature* . . .[32] (italics mine).

Locke also postulated that Hobbes' conception of humans in a state of nature did not include the positive aspects of reason. That is, for Locke, humans in a state of nature, along with their bestial desires and human passions, were possessed also of reason. If this is so, then human government could appeal to the rational side of human nature (as well as toward constraining the irrational side). The power of government need not be as strong as Hobbes believed in order to create order, because along with the sovereign power, a set of rational laws could be created, such that both the actions of citizens and those of the sovereign could be reasonably constrained. (Here we come full circle to Aristotle once again.)

For the rational basis of law, as derived from Locke, is twofold:

First, leaders, being human and potentially violent and brutal, must be constrained by rational law. The Hobbesian right of self-defence becomes extended to the right of the citizens to protection against all arbitrary government power. Again, such protection can only be established through legal authority.

Second, humans, possessed by reason, will not respond only to government force in terms of the maintenance of social order, but can also respond to the constraints of a set of laws, as long as these laws are rationally established and amendable, subject to debate and open to judicial appeal.

The fact that we know that political leaders may act violently and in their own behalf, and that we know that the population in general may act violently toward one another, demands that we establish legal constraints of such behaviour.

This is the rational basis of law. No 'natural law philosophy' needs to be invoked.

RATIONAL-LEGAL AUTHORITY IN THE MODERN WORLD

The rationality of law and legal authority transcends its origins in commercial contract law, and transcends its carrying classes, because it provides a rational basis for human political action beyond the structural and cultural realities from which it was derived. The more rationally oriented a population, the more likely such a population is to identify with the rationality of legal authority and to reject irrational, arbitrary or coercive political action.

However, legal authority may not transcend its origins if: (1) a new form of irrational legitimacy emerges which inhibits the modern rational perception of political action, or (2) technocratic-bureaucratic rationality replaces legal rationality as the dominant form of rational orientation to the world. (I have already contrasted these two competing forms of rational orientation to the world.)

Just as Locke, and Machiavelli[33] before him, warned us of the brutality and self-aggrandizement that political leaders might engage in without the constraint of law, so Weber warned of the despotic potential of political leadership within a massive institutionalized state and administrative structure lacking legal limitation and legal succession procedures.

However, the brutality and self-aggrandizement that can occur within modern bureaucracy is so much more concealed than that of Machiavelli's *Prince*. The public may be viewing a rational, stable, long-lived, productive administrative apparatus, and never viewing the executive suite any more than the peasants viewed the power-machinations of the Divine Kingly court. How many peasant or artisans knew of the struggles between eunuchs and harem-queens, 'mandarins' and nobles, for control of the throne?

The modern managerial elite of the state, economy, and military may also be able to conceal their byzantine activities while projecting 'public relations' images of stability and reliability to the populations embedded in the hierarchies of mass organizations.

On the other hand, modern populations know when they have no legal protection from the hierarchy and the elite; they know when they have no participatory input into decision-making; they know when they cannot recall their leadership and replace them with leaders sharing their own interests and desires.

Since the rationality of law is linked directly to the rational self-protection of a citizenry from its government, and to the rational limitation of the potential power excesses and selfishness of political leaders, it follows that if modern populations do feel threatened by their leadership and excluded from the decision-making process, and if modern populations perceive their leaders as ruthless and self-seeking, then rational legal authority could remain an important institutional structure under modern conditions.

Let me add, as an historical note, that the nineteenth- and early-twentieth-century Marxists failed to understand the significance of rational-legal authority. So concerned were they with establishing economic equality and non-oligarchic democracy that they tended to downgrade the importance of law. They made fun of 'law' as a 'bourgeois institution'. They made light of both constitutional and criminal law, believing they were just tools of the ruling business class. But, having denigrated legal authority, they failed to encourage its establishment in the foundling communist states. Lacking law, they soon discovered what it meant to live without its protection.

6 Maintaining the Middle-Class Majority on the High-Technology Industrial Capitalist Base

MAINTAINING THE MIDDLE-CLASS MAJORITY

A middle-class majority in and of itself is not sufficient for the engendering of, and maintenance of, democracy. The character of the upper class, the ideology and political culture, and the structure of the economy and the state, are equally important. However, Aristotle's theory of the link between a middle-class majority and a stable, relatively democratic, mixed polity,[1] still holds, in that the rise of a majority middle class remains one of the critical causal factors engendering democracy.

Given the existence of a large, prosperous middle class, the possibility for democracy may exist. Even within the context, for instance, of semifeudal Argentina,[2] the desire for democracy increased as the size of the middle class increased. And, even in Russian-dominated Eastern Europe, where the democratic tradition was weak (outside of Czechoslovakia), as the middle class became enlarged and became well educated (and as the working class became absorbed as a lower-middle-class portion of the middle class), the demand for democracy grew dramatically.[3]

In this treatise, we have attempted to show that the new middle strata of modern society may carry democratic elective affinities, and may, therefore, act as an Aristotelian mediating class, supporting and maintaining democratic processes of government.

If we accept, for the moment, that within modern high-technology industrial capitalist society – a social context far removed from the ancient polis – the growth of a majority middle class could be linked to the possibility of democracy – to its maintenance where it exists and to its establishment where it does not – then it follows that a society should do everything in its power to increase the size, prosperity, and educational level of its middle class.

Aristotle suggested a set of policy guidelines for the expansion of the middle-class majority, the raising up of the poor, and the encouragement of the rich to give of their wealth for the common good.[4] What would Aristotle's policy guidelines look like in the mass technocratic-bureaucratic context of modern society?

The policy guidelines which will be described are meant to be suggestive. What is hoped for is the promulgation of the idea of creating policies consistent with Aristotle's schema of class balance, wherein a majority middle class is fostered, nurtured, and expanded. Any policies which serve this end will be considered right and good, while any policies which result in the decline of the middle class, or which draw the lower classes toward tyranny and violence, or the upper classes toward oligarchy and repression, will be considered perverted.[5]

Of course, such policies may differ from one nation-state to another. In one society, fostering the growth of the middle class may be the key factor, in another upgrading the poor, in still another reducing the oligarchic tendencies of the rich. In England, for instance, fostering the expansion of the middle class is a major political problem – especially given the vast out-migration to Australia, Canada, and the USA. In Brazil, upgrading the poor is a serious necessity, if the polity is to become stabilized, while, in Mexico, curbing the oligarchic tradition of the upper class may be the central issue in establishing real democracy there.

Thus, each nation must develop Aristotelian policies which help engender the kind of class balance necessary for democratic stability. On the other hand, certain policies may have a general applicability because of the structural similarities of the emerging high-technology industrial societies.

CAREERS IN THE HIGH-TECH ECONOMY

The basic key to the prosperity, stability, and expansion of the modern middle class is the availability of stable, long-term career opportunities providing an adequate and comfortable income. Obviously, without a solid, well-functioning modern economy, a middle-class majority cannot be sustained.

The case of Russia in 1996 is a perfect example. The communist system generated a large, well-educated new middle class, but when this class – acting with the working class and the leader of the Communist Party himself (Gorbachev) – overthrew communism and attempted to establish a democratic government, they discovered that without the solid economic base of a functioning capitalist economy, political instability and extremism quickly emerged. The spectre of Weimar Germany began to appear as a haunting analogy. For, in Weimar Germany, the great depression and the war-indemnity payments crippled the economy, ruined the middle class, and set Germany on an extremist course beyond anything the world had previously known.

The Russians, are, of course, desperately attempting to establish a modern high-tech capitalist economy. However, it will be years before such an economy emerges. In the meantime, the condition of the middle class and the working class in Russia has deteriorated dangerously, leaving them open to the appeals of nationalist right-wing extremists and old-line communist hardliners.

Thus, an expanding high-tech capitalist economy is, obviously, a prerequisite for stable democracy, and the key factor in the sustenance and maintenance of the middle class.

INCOME AND THE MIDDLE CLASS

Along with a dynamic capitalist industrial economy, a modest, but adequate, income must be gained by the middle class. This seems obvious, but is essential. For instance, today in the USA, the income of the middle class looks similar on paper to that of their parents. However, this income equity has been maintained only because of the wives' entrance into the career world. The income of the 'dual-career couple' is excellent, but the effort here is double its forebears'. Furthermore, current outlays in the USA for child care (since the wife is not at home), education (nursery schools, prep schools, private colleges), health care, mortgage and other costs, have increased so dramatically, that the actual income has declined, while the effort to maintain it has increased.

Thus, in each nation, the 'actual' income of the middle class must be analysed, and an adequate, stable, income must be provided by the economy, and, by state policies which help to sustain this modest, but adequate, income in the face of fluctuating economic conditions. Economic adjustment and government programmes can be helpful – where they are needed – in creating and sustaining the stability of the middle classes. Such adjustments and policies are only needed where the economy itself fails to sustain the income of the middle class.

Throughout the high-tech world, such programmes as mortgage assistance, family assistance, college tuition assistance and universal health insurance have served to augment the income stability of the middle class. In the EU, Scandinavia and Canada, such programmes have been acknowledged, whereas in the USA, such programmes have been utilized effectively, but not acknowledged as such. GI mortgages and college tuition assistance were successfully utilized after the Second World War, for instance, while health insurance was offered by certain large corporations.

During periods of economic downturn, as in the USA today, Britain in the 1950s, Japan in the 1990s, and Germany with the absorption of Eastern Germany, conscious attempts – corporate and government – to sustain the income of the middle class must be made. During economic growth such policies can be phased out.

PROPERTY AND THE MIDDLE CLASS

Property in the modern context no longer means 'land' in the form of agricultural estates with bound labour. Property in the modern world means housing, cars, televisions and other technological products, plus stock property (rather than direct business ownership, in all but some small businesses).

From our Aristotelian theory of class balance, what one derives is the need to maintain a moderate but adequate property for the middle class. This has profound modern implications. For instance, in terms of home ownership, every effort should be made to make homes or apartments available to the middle class. This means that houses and apartments should be priced within the buying range of the middle class.

The trend – set in motion inadvertently by Reagan's de-control policies in the USA, and then spreading to Japan and Europe – toward building luxury homes, apartments, and condominiums, has been disastrous for the middle class – especially the children of the middle class – from New York to Tokyo. Such children, left out of the inheritance of a residence, may not be able to maintain their middle-class status. Such a situation is anathema to the long-term expansion and prosperity of the middle class.

STOCK PROPERTY

The middle class, from our perspective, should develop two strengths in regard to stock property: one, they must gain a moderate share of stock property, for their own economic prosperity, and, two, they must maintain the political ability to check the power excesses of the newly emergent financial institutions and corporate managers, who have gained control of, not only the stocks themselves, but the institutional functioning of the stock market.

First, general stock ownership of diverse investment stocks should be extended to a broader spectrum of the middle class. This latter is more a form of advanced 'banking'[6] (pension funds, mutual funds, and so on) than stock speculation. Second, ownership of company stock should be extended to middle-class employees, not just to top managers.[7]

In terms of diverse stock ownership, pension funds and mutual funds already provide such access to the stock market. However, at the time of writing, wealthy financial cliques, elite managers, and arbitragers have taken the lion's share of control (in American, European and Asian markets), while the small investor has gained little.

If new laws concerning pension funds and middle-class mutual funds were carefully written, the middle class could gain a greater share of monetary reward from this newly computerized and technologized stock market system.

In terms of the ownership of corporate stock by its employees, the ESOP (Employee Stock Ownership Program) was successfully institutionalized in many major American, German, and Scandinavian firms. Companies like Kodak, Uniroyal, Phillips and Volvo had gone far with the extension of stock to middle-sector employees. Sadly, the trend has reversed.

As things now stand, stock ownership is in the hands of six different groups: rich capitalist families; corporate managers; financial cliques (who have been raiding corporations, buying them out, breaking them up, and reconstituting them in a maze of non-logical conglomerate and oligopolistic entities); giant institutional investors, such as mutual funds, insurance companies, and pension funds; speculators and arbitragers, who own stocks only transitionally; and the middle class – largely the upper middle class, who own dispersed shares of stocks and bonds – with no controlling interest or financial import, but with only a hedge against inflation.

Obviously, the power of the financial cliques, managers and institutional investors needs to be re-regulated, while the share of stock-property allotted to the middle strata must be increased. Otherwise the dangerously de-stabilizing tendencies of the current system will increase. Kevin Phillips, in *Arrogant Capital*[8] has recently warned about 'electronic speculation' and the 'financialization of America'. We should heed his warning.

TAXATION AND THE MIDDLE CLASS MAJORITY

The middle classes provide society with the bulk of its revenues. Therefore, it is imperative that the middle class pay its fair share of taxes. However, taxes on the middle class should always be set at a moderate rate, because the prosperity and security of the middle class are at the core of democratic stability.

Now, we shall discuss the Scandinavian programme of high taxation and very complete free services for the middle class, shortly. Scandinavia has been very stable and very democratic politically. For the moment, let us stay with generalities.

The taxation programmes of most modern societies were originally set as 'progressive' (under the pressure of the socialist movement) – that is, heavy on the rich, moderate on the middle class, and light on the poor. This system was correct – in Aristotelian terms – in its inception, and it is still correct today. For, progressive taxation tilts towards the middle class in terms of wealth accumulation, and away from extremes of riches and poverty.

However, this progressive taxation system – from the USA to Great Britain to Europe to Japan – came under siege by the wealthier citizens, who sought to lower their tax burden. This is only natural and certainly politically expectable. Unfortunately, however, the rich succeeded too well in the USA in creating loopholes and exemptions for themselves. As one would expect, with the decline of tax revenues from the rich, the taxes on the middle class increased. This process slowly occurred between 1950 and 1960, was stopped by Kennedy and Johnson, but then was irresponsibly unleashed during the Reagan years.[9]

The correct ratio for the taxation system should emerge from the desired direction of proper class balance, and, of course, it should serve to improve the market economy, not impair it.

TAXATION IN SOCIAL-DEMOCRATIC 'WELFARE STATE' NATIONS

Taxation in the countries where social-democratic parties have had a major policy impact is of a different order than in the more purely capitalist countries. First in Sweden, Denmark and Norway, then in Germany, and finally in most of the European Union nations, a relatively high rate of taxation is combined with a wide range of free services provided for all citizens.

The middle classes and the rich are heavily taxed in order to provide for free health care, free education, free universities, free nursery schools (and now free day-care), and other services, such as pension funds and job-training programmes.

Now, the middle classes are taxed too heavily from both an Aristotelian and Keynesian perspective. However, since the return of excellent services has been achieved in all of these nations, the stability and prosperity of the middle classes has been assured by the excellent quality of these free services. Therefore, though less discretionary income is allowed for the middle classes, and this reduces both their 'propensity to consume',[10] and their ability to invest in stocks or businesses, nonetheless the stability of the middle classes is increased. The stability is increased because no health crisis can ruin a family, educational costs cannot drain income away – even the loss of a job will be cushioned by government support in retraining and placement in a new job.

Generally speaking, then, in terms of democratic stability, the social-democratic trade-off of cradle-to-grave welfare state services in return for high taxes works well.

However, there are negatives: in terms of entrepreneurial expansion, the lack of discretionary capital is definitely problematic, and consumption levels of new products is somewhat inhibited, due to the higher taxation.

The key fact is this: as long as the middle classes receive excellent services for their tax dollars, the tradeoff is viewed as right and good. If these services were to deteriorate, or if taxes simply rise too high, then a tax revolt would be possible.

TAXATION ON THE RICH IN WELFARE STATES

If taxation of the middle class is high in welfare-state nations, taxation on the rich is very high. Seventy to 80 per cent of the income of the rich may be taken in taxation.

The heavy taxation of the rich in social-democratic nations is called 'transfer payments'.[11] That is, the wealth of the rich is transferred to help fund the social services provided for everyone (including the rich) in these societies.

Is heavy taxation of the rich a good or a bad idea? In terms of economics, in the Keynesian formulation, capital taken from the rich, if invested by the government into industry, R&D, or infrastructural development is a good thing. This, according to Keynes, is good industrial policy, because in capitalist-industrial societies, the rich tend toward stock market and rentier investments rather than toward industrial investments.[12] If this is accurate, then taxing the rich and investing their money in the industrial system and the human infrastructure is good.

But, here we are discussing democratic stability. Is it good to heavily tax the rich?

Here is the good: taxing the rich heavily creates revenue for transfer payments to the middle and lower classes. These transfer payments help fund the health care, educational, job training, and other services which are available to all classes in society. Thus, they help stabilize the middle class.

Further, these transfer payments from the rich also help fund programmes for the working class and the poor. Job-training, and retraining programmes, skills upgrading programmes, family support programmes – such as the family allowance, day care, and nursery school programmes – all help raise up the poor and lift the working class to lower-middle-class status. Both these processes are very good for democratic stability. They militate towards a class balance wherein a middle class abounds, and wherein the rich and poor are not extremely separated from the middle class in income or life-style.

But, what do the rich get out of all this in welfare state societies? Why do they go along with policies that directly take wealth from them? Here we must look at the specific cases in question.

In Scandinavia, there is a kind of 'tribal-paternalistic' cultural orientation. That is, rich individuals and families feel that it is their duty to give back to the total society, and share their good fortune with the others. The case of Carlsberg Beer in Denmark is illustrative here. The Carlsberg family, having made a fortune from their beer company, willed their profits to the Danish people – forever – to fund education, health care, and so on.

Few American entrepreneurs would have done what the Carlsbergs did, but other Scandinavians have acted similarly.

In Germany, the motivation of the rich has been different. First, out of fear of a socialist revolution, Bismarck, and later industrialists and aristocrats, created a welfare state to coopt and pre-empt a socialist revolt. Having created it, they were willing to fund it, for their fear of their immanent overthrow was greater than their anger about being over-taxed.

After the Second World War, guilt and revulsion over the Nazi atrocities, and at their complicity in them and lack of resistance against them, motivated the post-war German rich and aristocrats to continue to improve the welfare state services to all Germans (and to compensate the German Jews).

What about now? Why should the European Union rich allow themselves to be overtaxed (by capitalist standards in the USA)?

The answer is that the EU welfare state transfers payments to industry as well as to the general population. That is, in all the EU countries, the government actively subsidizes industrial production. Established industries are encouraged – such as chemical industries in Germany or the tyre industry in France. And, fledgling or new-area companies, such as Airbus or Phillips Electronics, or the new computer consortium, are all heavily subsidized to help them compete against American and Japanese firms better in these new areas.

Further, research and development are also heavily subsidized by the government, as are state-of-the-art infrastructural improvements, like French high-speed trains and German roadways, along with telephone and computer link-ups.

Thus, the industrial rich, though heavily taxed, also receive excellent services in return for their tax dollars. The 'transfer payments' in this case, go from the individually rich to the industrial systems in the larger sense. This kind of 'transfer' is exactly what Keynes insisted was good for the modern capitalist industrial economy.

In terms of democratic stability, the rich do not counter-revolt, because it is in their economic interest to expand and modernize the industrial system.

Again, in the USA, the situation is perceived quite differently, because government investment into industry, R&D, and infrastructure has always been shrouded in military garb. It was accomplished, but no one admitted it was done. The rich and the academic economists insisted – and still insist – that a free market system, fully *laissez faire*, was the reality. This is false. However, believing that it is true leads the American rich toward a withholding of taxes, a denial of transfers to the middle and lower classes, and, to de-controlled capitalist activities.

The lack of taxation of the rich in the USA has created both a lack of 'transfer funds' for good services, and an overemphasis on the financial end of the economy.

RAISING UP THE WORKING CLASS TO LOWER MIDDLE CLASS STATUS

One of the keys to expanding the middle class in modern societies is converting the working class into a lower middle class, that is, by raising the level of education of the workers and by bringing their wage level up. The workers begin to be able to buy houses, apartments, cars and appliances and to act intelligently as citizens at a level near that of the middle class. What is necessary in all high-technology capitalist industrial societies is to treat the working class as a lower middle class. That is, academic education and technical-skill education must be made readily available to the workers – both male and female. And wages and benefits should be maintained at high enough levels so that this newly created lower middle class can gain at least a modicum of security and a sufficient consumer lifestyle.

Now these programmes and processes can be played out differently in different nations. But the result must be the same: the establishment of a solid lower-middle-class addition to the middle class.

TECHNICAL AND SKILLS TRAINING

In order to insure that the new working class connected to the high-technology global economy attains high wage and benefit levels, it is necessary to provide them with the exceptional training needed. That is, today's worker – though he or she may begin in sweat-shop factories – will need to attain assembly-line skills. These skills will include automation and computer familiarity.

To be sure, old-fashioned sweat-shop factories still abound, in terms of the assembly and packing segments of modern manufacturing. But even in the developing nations, high-tech skills are becoming necessary. The new economy has spawned new kinds of factories. These new factories are computerized, robotized, and are often very technical in their assembly-line processes. Therefore, both blue-collar and white-collar workers need advanced technical skills training.

The Germans and the Japanese have pioneered the most advanced systems of worker technical training. The Germans do it through the creation of very good technical high schools, while the Japanese do it through excellent on-the-job training programmes and skills-up-grade programmes.

The German technical high school system is so good that it should be copied, especially in Eastern Europe and the former Soviet countries, whose educational systems are already like the German. However, there is a flaw in the German system. That is, it 'tracks' children early into technical, as opposed to academic education. This inhibits the upward mobility of the working class, and, from our point of view, this is anti-democratic. On the other hand, the skills training is so good that the German economy runs very well, the workers adding in-put and perfection to industrial products.

The solution to this dilemma is a process the Germans are already providing. The children who fail to pass the academic track examination are now able to take it over and over, as often as they like. This, combined with some aggressive tutoring made available, could democratize the system more fully. Mobility to middle-class business and professional careers would then become more likely for the children of the workers. From our point of view, expanding the middle classes in this way is a stabilizing phenomenon.

Now, whether the children of guest-workers should gain this kind of mobility – and gain full political citizenship – is another matter which the Germans will have to face, as guest-workers like the Turks, Arabs, Africans, and now, Eastern Europeans, no longer wish to return home. In this latter situation, ethnic conflict, rather than class conflict, threatens to destabilize Germany and nations using the German system.

The Japanese system of technical training is equally as effective as the German. And its strengths and flaws are the same. The Japanese workers are among the most skilled in the world, and contribute to product excellence. But Japanese workers gain upward mobility more slowly than they could because they are tracked out of the college line early on. The Japanese have provided more chances for those children who fail the academic exam, but much less than they could. The Japanese should provide more help for the academic examinations to the working-class children. But they do provide excellent and rigorous grade school and high school education.

Finally, the Japanese also have a problem with multiculturalism. They deny opportunities to Korean, Filipino, and other guest-workers living in their midst. This could become a problem if the number of such guest-workers increase, as they have in Germany.

In the United States, the greatest failure has been in the area of technical skills training. Our colleges are excellent, but our technical schools are a joke. In high school, little or nothing of value for job skills is taught. Thus, if a student does not go to college, the high-school experience becomes meaningless. The 'mid-kids', who are not academically motivated, and who are not learning-disabled, get no special attention or training at all!

Technical two-year colleges and institutes are popping up everywhere – and some of them are good, but most have been recently surveyed as quite poor.

USA workers are hard-working and well-motivated, and, they are intelligent, but are often ill-trained, lack computer skills, and are ignored in the productive process. In a few companies – such as Saturn and Kodak – Japanese skills training and worker involvement have been adopted. But in most American companies, the nineteenth-century attitude that workers are expendable is adhered to.

Obviously, the USA needs a system of skills and technical training. It should probably begin at the high-school level, for non-college-bound students. But, it should be offered to high-school graduates – and even college graduates – who seek industrial and office jobs where such skills have become indispensable.

Without this latter system, downward mobility could occur, with the children of the working class becoming less prosperous than their parents, instead of moving into the middle class. The fear of downward mobility has – in part – generated new political movements, which reject the Demo-crats and Republicans and are searching for candidates who will talk to the needs of those suffering downward mobility, or at least the fear of down-ward mobility. Candidates such as Ross Perot and Pat Buchanan have recently surfaced, and others will appear until this problem is solved. And the right-wing militia groups are usually made up of working-class men who have no hope of a real middle-class life-style.

Further, and tragically, the 'mid-kids' who do poorly in high school and receive no skills training often turn to drink, drugs, vandalism, violence and rebelliousness. Violence, teen pregnancy, and heavy drug-use eman-ate from the high-school kids who feel they are failures. There are other causes, such as the high divorce rate and lack of community. Nonetheless, programmes of skill training and job placement could help the children of the working class, and all other adolescents who are not college-bound.

POLICIES CONCERNING THE POOR IN ADVANCED INDUSTRIAL SOCIETIES

Introduction

Aristotle said:

> measures should . . . be taken to improve the lot of the common people by a system of social services, both public and private.
>
> It is the habit of demagogues to distribute any surplus among the people; and the people in the act of taking, ask for the same again. To help the poor in this way is to fill a leaky jar . . . yet it is the duty of a genuine democrat to see to it that the masses are not excessively poor.[13]

> Poverty is the cause of the defects of democracy. That is the reason why measures should be taken to ensure a permanent level of prosperity. This is in the interest of all classes, including the prosperous them-selves; and, therefore, the proper policy is to accumulate any surplus revenue in a fund, and then to distribute this fund in block grants to the poor. The ideal method of distribution, if a sufficient fund can be accu-mulated, is to make such grants sufficient for the purchase of a plot of land; failing that, they should be large enough to start men in commerce or agriculture.[14]

> If such grants cannot be made to all the poor simultaneously, they should be distributed successively, by tribes or other division: and, meanwhile,

the rich should contribute a sum sufficient to provide the poor with payment for their attendance at the obligatory meeting of the assembly . . . (and the law courts).[15]

We must make a distinction in the modern context between two different classes of poor. First, there are the poor emanating from the declining industrial working class. They are the unemployed workers who cannot be re-employed within the factory system. Second, there are those who have never been employed as industrial workers. They are either immigrants who have come into an industrial country only to find that the new automated economy has no room for them, or unskilled, marginal seasonally employed labourers who never quite made it into the factory system or union-organized jobs. Among this second class, the permanently unemployed or marginally-seasonally employed, another distinction must be made. This underclass must be divided into two portions again. First, there are the working poor. They are part of the underclass, but can be easily absorbed into the working class. Second, there are the unemployables: those members of the underclass who have 'retreated' into drugs or drink or those who have taken up deviant careers in crime, vice, gambling, petty crime or anomic violence. Some members of this portion of the underclass may be absorbable into mainstream society, some may not be.

Each nation-state has a different set of problems regarding the underclass; however, in terms of the displaced working class, the problems are becoming similar in most advanced industrial nations.

With the new trend of moving corporate production to the Pacific Rim, Mexico, and other Third World nations, the problems of the loss of jobs for workers in the 'First World' nations will increase dramatically. Factories are being closed down all over the United States and Europe, and moved to the Third World.

Therefore, as we have emphasized, programmes within the advanced industrial nations for the re-training and re-absorption of workers must be speeded-up and improved. Furthermore, the educational programmes for the children of the working class also need urgent improvement.

Policies for the Underclass of Poor

The underclass of poor are a new class linked to the new technocratic-bureaucratic economy. They are a new class in the sense that the un-educated cannot be absorbed into the technocratic-bureaucratic economy the way they could be absorbed into the industrial economy. Most modern societies will continue to have large numbers of individuals who are educationally disadvantaged. Different nations, however, now have, and will continue to have, very different situations in this regard.

For instance, the situation of the United States is atypically difficult because of the mass migration of rural southern blacks to the northern industrial cities, and because of the in-migration of millions of poor, uneducated Latin Americans. At first, the Blacks (and Latins) were kept out of the industrial economy because of racial prejudice, but by now, the problem is structural. The rapidly automated economy simply doesn't need labourers anymore.

Thus, the problem in the United States is a massive problem in terms of the underclass.

In Germany and Scandinavia, the underclass is foreign. The guest-workers from Turkey, Greece, Italy, Yugoslavia, etc., are invited as an industrial working class, but when they are not needed, they are sent home. The problem of the underclass is thus avoided in these nations, but the problem is then thrown back to the nation of origin of the guest-workers wherein the underclass continues to exist.

In Britain, the in-migration of Irish, West Indians, Indians, and Pakistanis, along with the continuing unemployment problem among the British working-class whites, has perpetuated a long-term underclass problem there.

Third World nations, though industrializing rapidly, as say in Brazil, are experiencing an underclass problem of monumental proportions. Usually, the underclass far outnumber the tiny working class and the traditional peasantry (from whom the underclass are drawn as traditional-agrarian society breaks down). This would be true in the Philippines as well.

Obviously, policies for the absorption of the underclass of poor would have to be carefully tailored to the economic and political circumstances of the particular nation in question.

In a general sense, Aristotelian-oriented policies concerning the new poor should follow certain basic principles. Simply giving money to the poor, is 'like pouring water in a leaky jar'. Providing careers and education become the keys to the absorption of the underclass into the working class, and the lower middle class. Let us look at these in reverse order.

Educational enhancement.   A certain small percentage of the underclass can, with educational programmes, achieve new middle-class status. Further, a larger percentage can, through educational programmes, achieve at least working-class status. The educational support programmes would have to begin almost from babyhood, and be continued through to the college level in order to be fully effective. Such programmes include: the reading of stories to poor children; preschool, head start, enrichment, and tutorial programmes; first-grade to high-school tutorial, literacy enrichment, cultural enrichment ('higher horizons'), etc. Such programmes as the high school 'Upward Bound' summer programme, along with cultural enrichment and prep-school programmes, would have to be continued.

In college, special study, reading, maths, non-credit remedial courses and tutorial work would have to be increased.

All this effort would only upgrade a small percentage of the underclass – and even then they might not function on a par with their middle- or working-class college graduate contemporaries. The majority of the underclass may not benefit from the educational programmes for three or four generations.

What policies should be instituted for those members of the underclass who do not benefit from the educational programmes?

Job Programmes.   The underclass cannot be absorbed as industrial workers in a technocratic-bureaucratic economy. If they do not make it into the economy through education, then they must be absorbed through temporary, but long-term, job programmes. The job programmes will probably have to be created by the government, because the corporations are automating, and do not need any unskilled labour.

However, we are not advocating make-work jobs. There are jobs that need to be done in modern society that are not being done. For instance, in the American cities, the level of services has deteriorated to a frighteningly low level. The cities need police, fire, sanitation, mass transit, parks and health care workers. These jobs are hardly make-work jobs. As the children of the working class are absorbed through education into the new middle class, the children of the underclass can be absorbed into these urban service careers.

In rural areas, there are also jobs to be done that are not being done. Anyone who has visited America's small towns and countryside knows that the small towns have deteriorated as badly as the cities. Many of the houses and stores are falling apart. In the countryside itself there are shack areas, while the woods are filled with beer cans and rusting auto hulks. Thus, rural jobs could exist as well.

Finally, for the underclass individuals who cannot be fitted into urban or rural career-line jobs, a temporary, but long-term, job-corps, or WPA programme should probably be established.

Where would the money come from? The money is already being given to the poor, but in the form of welfare payments, and crime payments. Welfare money is money thrown away. The poor 'in the act of taking, must then ask for more'. Crimes of theft and violence are at a very high level. Auto-theft, home burglary, bike theft and mugging are routinely accepted in our society. The middle class pays for both welfare and crime.

Wouldn't it be better to subsidize a job and career programme than a welfare and prison programme? And, shouldn't the rich and the corporation pay their fair share? Further, service jobs would be created for the middle class, and the youth of the middle class could serve as volunteers, or summer and part-time paid workers in the effort to absorb and rehabilitate the underclass. All of these latter would be beneficial to class balance and stability, whereas the welfare and crime systems set the middle class and underclass against each other.

For those sceptical about what is possible and what is utopian, I can recount the attitude of the middle class toward the working class and their college potential in the 1950s. Only a small percentage of working-class children went to college then. Most were high-school dropouts or 'D' students. They quit, fought in street gangs, were filled with hostility, joined the Army and then became blue-collar workers (the women became either low-level secretary-clerks, or blue-collar workers, or married early with early pregnancies). Yet, twenty years later, a large percentage of the children of the working class are completing college. They are still more hostile, less well-prepared educationally and attitudinally, than the middle class, but are making it through college and are successfully retaining the lower-level new middle-class careers.

The situation of the underclass would be more difficult. Their social and family condition is much worse than that of the working class. The number of unrehabilitable and unemployable is far greater. But the long-term results would parallel those of the working class.

Drawing the Rich towards the Middle Class

From the Aristotelian perspective, the rich can be too rich, showing no respect for the law or the community.[16]

If the rich are too rich, according to Aristotle, they will 'draw the constitution away from democracy and towards oligarchy' (plutocracy).[17] Our thesis focuses on the expansion of the middle class, and on devices for insuring that the rich 'give of their fortunes for the common good'.[18]

TAXATION OF THE RICH IN HIGH-TECH CAPITALIST SOCIETIES

An Aristotelian tax policy would centre on a fair mechanism of taxation that would utilize the tax revenues from the rich both for transfers to the middle and lower classes and investments in the industrial economy.

The progressive income tax itself, of course, was supposed to be the mechanism to facilitate transfers from the rich to the rest of society. However, over the years, as mentioned, in the USA the rich have been powerful enough to create loopholes so extensive that their taxation level has been substantially reduced. With this reduction in taxes, the transfers to the industrial economy and to other classes have also been substantially reduced.[19] It is imperative, therefore, that in the USA a proper level of taxation of the rich is resumed.

The loopholes should be closed. Or, if this is too complicated a process, then a high minimum tax should be levied on all individuals (and corporations) with an income above a million dollars. This would bring our taxation system closer to that of the Europeans and Japanese, and allow for our social service programmes to be properly expanded, rather than being cut, as is presently (1996) the case. Investment for R&D, infrastructure, and industrial technology would also become available again.

Taxation of the rich should also take another tack. That is, a luxury tax should be instituted. Items like $10 000 watches, $40 000 cars, yachts, jewellery, and other highly priced extravagances should be taxed. Though such trend-setting 'life-style' consumer items are exciting and lead to broadened consumer choices, as Hayek asserted,[20] if not taxed and limited to a degree, they tend to drive up the prices of the standard consumer items, making it more difficult for the middle class to purchase the more modest lines of consumer goods. For, if a $10 000 watch sells too easily to the rich, then $100 watches tend to move up in price to $200 and $300 – often incorporating unnecessary features, or simply, gaining price through snob appeal. This is true of cars as well, wherein the inclusion of luxury features, like leather and turbo chargers, all-wheel drive, and complex stereos has driven the price so high, that middle-class consumers must lease these cars – for they cannot afford to own them.

Thus, fancy, trend-setting products are good for economic growth, technology advancement, and social excitement, but the prices must be contained or they lead to the ruin of the middle class.

CORPORATE TAXATION

Taxation of the corporations is a different issue. Here industrial expansion, technological innovation, and economic growth are the real issues. Any taxation policy that encourages industrial production is good. Any policy that inhibits or diverts it is bad.

In Europe and Japan the corporations are taxed fairly, and the government invests carefully in their growth. In the USA, corporate taxation policy is a complex web of ineffective mechanisms. The American corporate system worked smoothly, in part, because of military investments in technology, infrastructure, and production. The American system – with military investment dwindling – is now in crisis.

THE TAXATION OF STOCK TRANSACTIONS

We have discussed the high-tech revolution in the capital markets. We have also negatively analysed the Reagan decontrol policies surrounding the markets and the banks. So much capital is changing hands through the various exchanges, and it is so highly computerized, that breathtaking fortunes can change hands in seconds. As we write, the Chicago commodities exchange has added speculative possibilities in 'junk' – that is, recyclable commodities, such as glass, plastic, paper and metal.

Somewhere within this new global system, accountability to one's nation – or to the non-speculating populations in general – must be established through an increased taxation programme. This could be programmed into the computerized systems, taxing transactions and excessive profits in a rational way.

AN INCOMES POLICY FOR CORPORATE AND FINANCIAL MANAGERS

As the corporate and fiscal managers become powerful new upper strata in modern society, it becomes imperative that they – like the capitalist-industrial rich – give back some portion of their newfound wealth to the societies in which they function. Taxation policies for upper-income individuals have already been discussed. However, in the case of managers, we are not dealing with profit-generated income; we are dealing with salary plus privileges: perks, such as stock options, bonuses, travel and entertainment expenses, golden parachutes, etc. Given the complex income sources, taxation policy is insufficient. An income limitation policy may be necessary.

In Japan, the performance level of managerial personnel is very high, even though their relative income *vis-à-vis* their American counterparts is very low. In the USA, corporate managers have been routinely taking huge bonuses, while corporate profits have declined and hundreds of corporate employees have been fired. How can democratic stability be maintained in a situation where hundreds of thousands of middle managers and blue-collar workers lose their jobs or take lower salaries and benefits, while at the same time the managerial elite grow richer?

The Germans, like the Japanese, have a fair managerial salary and perk level. Americans would do well to study the European and Japanese income levels for top managers, and to limit the salaries and perks of corporate and financial upper management. Both the economic and political power of the managerial class has become extensive – beyond the point of Aristotelian safety.

EDUCATIONAL PROGRAMMES FOR THE RICH IN HIGH-TECH SOCIETIES

The rich – business, financial, or industrial – are no longer the intellectual elite of society. Neither the British gentry nor the German Junkers produce the scientists, specialists or scholars who make the great intellectual break-throughs that move modern knowledge forward. The intellectual elite is now drawn from the middle classes.

Therefore, why not enrich the educational programmes of the rich? The American private schools, the British boarding schools, and elite schools worldwide, should emphasize science, maths, and computer programmes for the rich students. Of course, they should retain their excellent literature and language programmes as well.

Further, in order to encourage the children of the rich to contribute usefully to modern society, merit scholarship to elite colleges should be awarded to them where they qualify. These merit awards should be granted with honours, rather than money, the idea being to reward excellence in scholarship with high status. Specific merit scholarships in science, maths, and technology should also be awarded in the same way.

These educational support programmes are important, because the children of the rich – from New York, to London, to Singapore – often become dilettantes, rather than professionals. And worse, they have been squandering their fortunes on luxury goods, and worse yet, distorting the electoral process by buying television candidacies through huge campaign expenditures.

The encouragement of serious adult careers for children of the rich could be accomplished by the introduction of the special educational support programmes in the elite private schools.

CONCLUSIONS ON THE MAJORITY MIDDLE CLASS AND DEMOCRACY

In order to remain stable, the modern democratic polity, like the mixed polity of antiquity, must continuously create measures which work to create the proper class balance. The policies that help to produce an ever-larger and more prosperous middle class must constantly be improved. The policies which help to upgrade the position of the poor and integrate them into the middle class must be ever bettered, and the policies which encourage the rich to give of their wealth for the improved prosperity of all classes must be reevaluated and reinvigorated continuously.

Let me repeat that Aristotle's formulation of middle-class democracy, or the mixed polity, should be seen as a process, rather than as a static phenomenon. The process is in constant need of balancing. Any major imbalances in the system can draw the constitution towards oligarchy, tyranny, or years of inconclusive anomic violence.

In each nation, the processes and programmes for the creation of, and, sustenance of the middle classes, is different. Therefore, each nation must design Aristotelian policies which will help create a solid middle class majority. In this way, democracy will be stabilized.

Of course, this process presupposes the establishment and expansion of a solid high-technology industrial capitalist economy to undergird and support the middle-class majority, and the establishment of legal-democratic political institutions to allow that majority to govern itself.

Part III

The Empire Analogy: Bureaucracy Against Democracy

7 Bureaucracy as a Despotic System of Domination

Within modern societies, wherein legal-rational authority is institutionalized, bureaucratic administration has emerged, and civil liberties have not disappeared. Therefore, why doubt the future in this respect?

The answer is that bureaucratic organizations have shown increasing resistance to legal constraints, and have exhibited a growing tendency to hold to their own rules and regulations even where these have been explicitly forbidden by the courts. Further, few activities of modern public life have remained outside of bureaucratic organization. Therefore, one cannot be sure, at this early phase in the history of legal-bureaucratic authority, that such a fusion of conflicting legitimation processes can become viable.

We have already contrasted the kinds of rationality typical of legal authority and bureaucratic administration. We also suggested that though bureaucratic organizations exhibit administrative rationality, their political processes lean towards irrational mechanisms of consent-getting. These irrational mechanisms of legitimation are so ingrained in the organizational structure of bureaucracy that they will conflict with the rational processes of legitimation institutionalized within legal authority. With the extension and growth of bureaucratization into so many new spheres of public social action, these irrational processes of legitimation could become the dominant form of consent-getting within modern societies.

Let us look specifically at the irrational processes (or mechanisms) of legitimation which typify modern bureaucracies.

DEPERSONALIZED CHARISMA

In the ancient empires, such as Egypt, China, and Persia, depersonalized and manufactured charisma[1] were brought to heady heights: clan charisma became blood aristocracy wherein individuals from royal clans monopolized all leadership offices; knights and priests became royal classes linked to royal clans; the office of kingship became surrounded by such fantastic manufactured charismatic effects that it came to be conceived of as divine; kings wore the most magnificent clothing, held magical sceptres and swords, sat on magnificent thrones, wore crowns and jewels, were surrounded with armies of knights, and had harems of wives for their pleasure.

The court and the pomp surrounding the kings created a manufactured charismatic aura unsurpassed in history. The office charisma of the kingship invested its incumbent with awesome charisma, though the incumbent might actually be incompetent, retarded, psychotic, a child, or a woman (in a male-dominant society).[2]

How does depersonalized and manufactured charisma exhibit itself in modern bureaucracies?

In place of clan charisma, there is organizational charisma. That is, giant organizations – private or public – attempt to develop charisma of their own, such that any individual who works for that organization shares in the halo that the organization exudes.

Public relations departments, advertising departments, and personnel departments work specifically to create such images, using every modern manipulative technique at their disposal. Government bureaus, government agencies, and private corporations in most modern nations engage in such activities. Even the Pentagon advertises extensively, and makes public 'parks' out of its discarded aircraft carriers, submarines, and bombers.

Furthermore, modern bureaucracies exhibit a hierarchy of carefully defined offices. The incumbents of these offices are invested with a certain amount of authority simply by their occupancy of the particular office in question. Now, sociologists have done many studies showing that there may be certain 'natural' leaders, or old-timers to whom people defer, or approach, when they want to get things done. The 'informal' leadership structure exists,[3] and I do not wish to down-play its effect on bureaucratic operations.

The hierarchical authority structure of bureaucracies does, however, endow the incumbents of superior offices with a kind of depersonalized office charisma. If the incumbent of such an office is minimally competent and holds such an office for a fair amount of time, that person may gain a certain amount of charismatic aura.

Middle, and near-top managers, bureau heads, executive directors, captains and colonels, bishops and cardinals will gain authority beyond their personal attributes through office incumbency. The 'natural' leaders and old-timers may resent them, but their power and the deference paid to their power is backed by the hierarchy and accepted by the line workers in the chain of command.

The point is that a manufactured charismatic aura comes to surround the elite bureaucrats – especially in terms of their image to the middle and lower managers. In fact, we find a remarkable paradox here. That is, on the one hand, bureaucrats attempt to remain faceless and anonymous, avoiding the kind of publicity that electoral politicians and swashbuckling entrepreneurs bask in, while on the other hand, the elite managers of business and government are carefully surrounded by manufactured charismatic effects that cast a halo of omnipotence around them, making them seem to be superhuman.

The kings of the ancient empires made few public appearances, and their palaces were always behind sacred and forbidden walls. The people knew little of the actual personality of the incumbent individual; they knew only the outward trappings of the office of kingship. The more distant they felt from the king, the more likely the manufactured charismatic effects would work, and the aura of the office of kingship would be enhanced.

The situation of secrecy and facelessness of modern elite bureaucrats is similar. The public and the hierarchy know little about most of them, and see them only as surrounded by important people, wearing elegant clothing, chauffeured about in the best vehicles, frequenting the 'best' places.

This situation is egalitarian in that no aristocratic or oligarchic family background are necessary for either succession to elite offices or acceptance into the world of celebrities. However, though egalitarian, it is not democratic. The world of the elite is another world, as distant from the average individual or the middle managers as the forbidden cities of the ancient empires were to the peasants and artisans.

CO-OPTATION AND MODERN BUREAUCRACIES

Ancient kingly-bureaucracies successfully co-opted large numbers of individuals into their hierarchical structure. They drew in the brightest sons of peasants, the merchants, and the sons of the bureaucrats themselves. In this way, the revolutionary potential of the peasants, the merchants, and the bureaucrats was effectively obliterated. Upward mobility in the bureaucracy was egalitarian. The ambitious among the excluded classes could attempt to gain entrance to power and prestige through bureaucratic service. This they did, and this kind of co-optation did serve to enhance the legitimacy of the kingly regime.[4]

In the modern world, the co-optation will reach heights unknown in the ancient empires. The majority of the population will undoubtedly be absorbed into one 'white-collar pyramid' or another. In a sense, the entire population of modern society is becoming co-opted into business, government, or service bureaucracies.[5] This could mean exactly what it meant in the ancient empires, i.e. that democratic-participatory or revolutionary activity will be diverted into non-political work within bureaucratic hierarchies – the reward of successful upward mobility compensating the ambitious or disgruntled individuals in place of democratic electoral success or revolutionary leadership.

The unsuccessful individuals are easily repressed in such a situation because they are made to appear as failures, cast out of the hierarchy because of disruptive behaviour or inadequate skills. Those within the hierarchy find it difficult to identify with natural or charismatic or egocentric leaders because from their vantage-point within the hierarchy, such individuals do appear as disruptive, and their behaviour does seem inappropriate. In the bureaucratic setting, they do not appear as great leaders, but rather as naive leaders, lacking in 'organizational skills' and unequal to the byzantine demand.

The Bismarcks and Kissingers, not the Roosevelts and the Kennedys are admired within this kind of system. Of course, for those outside of the system, i.e. not co-opted into it, charismatic leaders or demagogic leaders would still have an appeal. But we have suggested that a greater and greater proportion of modern individuals are becoming absorbed by the bureaucratic organizations of society.

On another level, the co-optation of the modern individual by bureaucratic organizations becomes even more successful through the extension of 'gifts' to the populace. Medical and dental plans, pensions and stock options, savings accounts, etc.: all these gifts by the government or corporate bureaucracies serve to co-opt the individual's loyalty and thus aid in gaining the individual's consent.

Where such benefit plans are poorly administered or inequitably distributed, they can become a source of de-legitimation, of course. Where they become increasingly comprehensive, well-administered, and more equally distributed (in terms of the acceptable minimum for those at the bottom and middle levels if not in terms of actual equality of distribution) they can become a powerful source of legitimation.

IDEOLOGICAL MANIPULATION BY MODERN BUREAUCRACIES

In the ancient empires, massive ideological manipulation emanated from the kingship and the aristocracy. Along with the manufactured charismatic effects (magnificent clothing, palaces, courts, harems, etc.), an ideology was promulgated which specified that the aristocracy was royal and the kingship divine, that some had the right to rule while others did not, and that this situation was right and good. The domination of the aristocracy and kingship and their usurpation of great quantities of wealth and privilege was effectively legitimated by religious and traditional ideas.[6]

The kingly-bureaucrats were involved in the ideological justification process of the ancient empires, but their role and their relative success or failure in this legitimation enterprise varied greatly from empire to empire. In China, the bureaucrats were remarkably successful in generating their own ideology and their own legitimation,[7] while in the Middle East they were far less successful.[8]

What of modern bureaucracy? What kind of ideological manipulation emanates from it in an attempt to create legitimation and avoid rational evaluation?

Modern bureaucracies attempt to create an ideology of technical superiority, administrative efficiency (over large-scale operations), and meritocracy in hiring and promotion procedures. And, just as the mandarin's claim was based on some modicum of reality, so, too, is the modern claim. Why then is there ideological manipulation involved? Because: (1) Bureaucracies may or may not hold a monopoly on technical superiority. Sometimes they are technologically conservative and non-innovative, pushing out those with creative technical talents, while at other times they may subordinate technical necessities to managerial ones. Thus, the technical superiority of the personnel of modern bureaucracies is sometimes grossly overrated. (2) The meritocracy is marred by: (a) class, status, and ethnic group interventions towards the monopolization of upper hierarchy positions, and (b) by the Machiavellian manoeuvring for promotion and domination within the hierarchy. This takes precedence over administrative skill or technical expertise in the promotion process. And further (c) organizational loyalty and elite-clique loyalty is also directly involved in the succession process to high leadership roles. (3) The administrative efficiency over large-scale services often degenerates into inefficiency, red-tape obstructions, or even corruption. It may not be true at all that giant bureaucracies are the most efficient administrative units for the delivery of mass services or overseeing of large-scale projects.

Therefore, the egalitarian, meritocratic, technical superiority and efficiency claims of the modern bureaucracies are only partially substantiated and would not stand up to a fully rational performance evaluation by the populace. Thus, an ideology of technocratic superiority, administrative efficiency, egalitarian selection, and meritocratic promotions emanates from the modern bureaucracies as an ideological manipulation – a process designed to gain consent from the populace through irrational means.

COERCION AS PART OF MODERN BUREAUCRACY

When despotic regimes fail to gain consent through the irrational processes of legitimation, they often turn to coercion in order to maintain
themselves in power. Furthermore, coercion itself can be utilized by despotic systems to create a kind of acquiescence similar to consent in that
an acceptance of the political system occurs within the populace.[9] Rational
consent is not given to the regime, of course, but the acquiescence attained
is reinforced so powerfully through mechanisms that create fear in the
population that the regime may become long-lived, and stable (social order
may not be disrupted by revolutionary or anarchic political activity against
such a regime).

Coercion can be separated into two processes: psychic terrorization and
direct physical violence. The two go together, of course, the terrorization
always being backed by the threats of force or occasional carefully displayed uses of physical force and violence.

Ancient empires thrived on psychic terrorization and displays of violence. The slightest offence against the king or the aristocracy or priesthood brought instant punishment and often horrible death. In modern
bureaucracies, what forms of coercion take place?

First of all, it must be stated again that the rational processes of legitimation are repressed actively by bureaucracies. No civil liberties exist for
an individual acting within the bureaucratic structure. Free speech, free
press, freedom to organize and assemble, debate of policy decisions, fair
hearings for rule infractions, etc., these do not exist within the bureaucratic
structure. One can take one's case outside that structure to the courts or
the legislature in societies wherein legal democracy exists. However, within
the bureaucratic structure, legal-democratic rights are not recognized.

How do bureaucratic organizations utilize coercion? They utilize psychic
terrorization rather extensively. Now, I do not mean this in the extreme
sense, as it was so brilliantly described by Hanna Arendt (*The Origins of
Totalitarianism*),[10] relating to the concentration camps of Nazi Germany.
Nor do I mean to equate such usage with the public displays of mass
human sacrifice that characterized (early) Egypt and the Aztec Empire.
Nevertheless, psychic terrorization can take more benign forms and still
produce the desired effect, i.e. the cowing of the populace through the
spread of fear and the reduction of citizens' participation, charisma, and
rebellious activities.

The psychic terrorization used by bureaucracies focuses on: (1) The
threat of firing, or, 'riffing', the threat of non-promotion, the threat of
transfer to a less interesting, less important, less well-paying office, the

threat of transfer to the boondocks, or failure to be transferred to the metropolis. (2) The blackballing of 'troublemakers' or 'whistle blowers', or disloyals, or charismatics, or union organizers, or political dissenters, such that the career of the individual is destroyed, not just the particular job – such individuals may find themselves unhireable anywhere within the career system they were trained for.[11] (3) Computer files, or dossiers kept on all employees – all industry employees, or all national employees. Corporate, government, and service bureaucracies may keep such files[12] such that an individual may be threatened with having an 'undesirable' reputation for life. (4) Organizational spying and surveillance – trade secrets, government secrets, also personal secrets – and the threat of exposing them – the FBI, under J. Edgar Hoover, used this threat for years as have some large corporations.[13] (5) Along with the above techniques, the elite and middle managers sometimes stimulate rumours, sometimes speak directly to their underling staff of the possibilities of negative happenings, such as schedules changed for the worse, fringe benefits which could be reduced, work load which might be increased, procedures which may be contemplated, or cuts in personnel or salary which may be immanent.

All of this kind of information creates anxiety in the personnel and this anxiety turns to fear if such performances are actualized from time to time. Such fear becomes psychic terror if it is cleverly manipulated, and such terror produces an acquiescent subjugation. This is not consent, but its effect is the same – order and stability are achieved. This kind of legitimation can be long-lived.

What of the direct use of physical force or violence? Bureaucracies are authoritarian in structure and ideology, and they may become fully despotic if not checked by external legal-democratic constraints. There are many examples wherein bureaucratic organizations have resorted to physical violence to attain their goals or to maintain internal control. Of course, if the bureaucracy is attached to the power of the state, or if the bureaucracy is specifically linked with military or police power, then the potential for the use of violence is increased, even in a legal-democratic nation.

The question here is whether bureaucracies as such tend to utilize violence, not whether bureaucracies which are attached to despotic governments are used by such governments to commit violence. We know that the officials of despotic regimes will use violence, but what of officials leading bureaucracies in legal-democratic societies?

Actually, bureaucracies are not likely to utilize direct physical violence, nor are they prone to utilize the threat of violence – torture, murder. Bureaucracies are organizationally rational, they thrive on stability and longevity. An atmosphere of direct physical violence brings with it the possibility

172   *The New Middle Class and Democracy*

of counter-violence by organized revolutionaries, or, anomically generated violence by social deviants. An atmosphere of violence could bring charismatic leaders to the fore. None of this is desired by modern rational-bureaucratic organizations. Even those attached to despotic states have found it more stabilizing to avoid direct violence wherever possible.[14]

Thus, even the psychic terrorization tends not to be based on the fear of physical violence and death, but rather on job stability, career chances, location, promotion, and so on.

Individuals working within bureaucratic organizations are not as uncontrollable as peasants and artisans of old. Within the hierarchy, the rules and regulations and authority structure exert a direct pressure on the individual for conformity. Rebellion in such a situation is not likely to be violent, and, therefore, the authoritarian control exerted is also not likely to be violent.

One should understand, however, the potential for the use of violence by the elite of bureaucracies, and the willingness of middle levels of managers to 'follow orders' even in the case of the use of violence. Elites of bureaucracies have shown themselves quite capable of using violence when it is deemed necessary, and middle managers have carried out such violence with few whistle blowers to expose them. Cases of indirect violence already abound. That is, service agency directors have allowed facilities for the retarded, the insane, and the elderly to reach conditions wherein those serviced have died. David Rothman describes such instances at Willowbrook, the home for the retarded (in New York in the 1980s),[15] while Andrew Stein made his political reputation by exposing the desperate situation in the nursing homes in New York City (in the 1970s).[16]

One case of direct violence should be noted as a warning. The role of the chief executive officers of American corporations in the violence that occurred during the ousting of Salvadore Allende in Chile shows that the top managers of corporations, when their organizations are threatened (in this case by nationalization) may act violently to protect them.[17]

Finally, the case of Adolph Eichmann stands as an ominous warning in that, at his defence, in Jerusalem, he claimed he was only following orders, and that if he did not carry out such orders, someone else would have.[18]

Lastly, the CIA manual found in Nicaragua in the 1980s which coldly and analytically – like a textbook – described how to go about murdering the political opposition, shows how a bureaucracy, when attached to state power, can become violent in an apparently 'rational' manner. The fact that the American CIA is usually headed by a respected member of the Protestant-business establishment,[19] such as George Bush or Allen Dulles, rather than by some bloodthirsty outlaw or henchman of a sadistic, dictatorial fiend, points up the problem of the potentiality for bureaucratic violence even more urgently. At a party in Washington, no one would be able to pick out the ex-CIA director, or current CIA middle managers from the other VIPs and government bureaucrats in the crowd. Nor has any negative stigma become attached to the elite managerial officers of the CIA – witness the success of George Bush and the lack of electoral critique against him on these grounds. In fact, his 'experience' in foreign affairs worked for him, not against him, in electoral politics. No guilt was attached to the office. The Eichmann warning becomes all the more worrisome, given these developments.

REIFIED POWER AND MODERN BUREAUCRACY

The reification of power was one of the most frightening results of the political systems of the ancient empires. The state was viewed as a 'thing' beyond the control of humans. The divine king, the forbidden city, the court, the aristocracy – this was the world of the superhumans. They lived behind a shroud of manufactured charismatic effects, while the populace at large lived in a world of alienation and degradation.

It is quite possible that the reification of power could emerge again, separating the world of the superhuman celebrities, stars, and power elite from the world of the everyday individual. This was C. Wright Mills' nightmare vision.

The giant size of bureaucratic organizations and their global character creates a reified image of them such that they appear to transcend the collection of individuals that make them up. The top managers of government, corporate and service bureaucracies seem far away, remote from the individual – beyond his or her control and 'above' the individual in a world of power and prestige. Specialized experts with knowledge beyond the ordinary help to enhance the reified image.

The mass-media world of celebrities and stars joins with the world of corporate managers and statesmen and politicians and seems like a fairy-tale world beyond the mortal realm – like a realm of dreams and nightmares. All the beautiful and powerful and knowledgeable people exist in one world, while the average individual lives in another; but the modern individual is allowed to peek-in on the reified world through the mass media – but as a dreamer views a dream – in it, but without control over it.

Just like the forbidden city and the court and the divine king and the royal aristocracy, the power elite and the VIPs and the celebrities and the jet-set and the exclusive retreats and resorts and clubs and board rooms become a separate world from the ordinary. Power becomes a 'thing' beyond the reach or control of the 'average' individual in mass-technocratic bureaucratic society.

Political alienation and feelings of powerlessness to effect the political process overwhelm the individual who perceives this situation as reified.

If the reification of power occurs in the modern world, then truly, the 'iron cage of serfdom' will close around us, and like the Felahene of ancient Egypt we will watch a succession of 'kings' ruling our destinies.

We would not have written this book, however, if we believed that the reification of power was inevitable or could not be neutralized. There are dynamic sources of de-reification occurring in the modern world along with the sources of reification. The sources of de-reification emanate from the same general processes that could produce the reification. We shall delineate the sources of de-reification shortly. Here, in this context, let us be aware of the fact that the 'empire analogy' is no mere pessimist's quirk. The real possibility that the world of top managers, VIPs and celebrities could become a world apart from the average individual, and through 'media-manufactured charisma'[20] appear fully reified, is very great indeed. The feeling of distance from the world of technical and managerial decision-making is already profound within modern populations and alienation from the electoral process often results from this.

THE ETHICS OF BUREAUCRACY AS NON-HUMANISTIC AND AMORAL

From C. Wright Mills' 'higher immorality'[21] to Robert Merton's 'Bureaucractic Personality'[22] the bureaucratic ethos has been described in negative terms. The reason for this is that organizational needs, efficiency and profit needs, and stability needs, are given primacy over human needs, community needs, and specific individual needs.

Further, because of the mass nature of the production and delivery systems of giant bureaucracies, individual problems, if they are unique, are never given consideration – only the typical case is covered by the rules of the giant organization; these rules tend to be rigid rather than flexible.

Bureaucracies tend towards amorality rather than immorality. Chemical companies will make napalm and penicillin – whatever sells profitably. Food companies will make whole grain cereals or junk food – whatever sells. GM will produce unsafe cars if the price–profit situation demands it,[23] while Ford will market safety cars if they think that will sell. Mental hospitals will over-drug and electrify patients if it quiets them and allows for easier control. Good and bad, right and wrong – these are not paramount considerations in bureaucratic policy-making.

The organization is personified – made human – while the humans are dehumanized. The life of the organization and loyalty to it comes before the lives of people and a humanistic connectedness to them. The danger involved in such an amoral ethos is overwhelming. Hannah Arendt's account of 'Eichmann in Jerusalem',[24] though controversial on many issues, was not controversial on one: That Eichmann defended himself by insisting that he was just 'following orders,' that if he did not carry out the orders to kill the Jews, someone else, perhaps more efficient and officious than he, would have anyway.

Mentioning the extreme case of Eichmann is important, because it highlights the same attitudes held by the little bureaucrats in the little cases.

MERITOCRACY AS A NEW MANDARIN IDEOLOGY

Bureaucracy is not meritocracy. Merit is not the principle criterion for promotion in the hierarchy or for elite control in bureaucratic organization. Loyalty to the organization, a workaholic commitment to it, an ability at byzantine manoeuvring within the hierarchy and with the board, and, Machiavellian ruthlessness in terms of competition for the top spots, come before administrative or technocratic merit in the movement for advancement and control of giant organizations. Administrative success and technocratic genius are rewarded, and, therefore, merit is involved, but this is never all that is involved.

The 'Power Elite' or the 'New Mandarins' could use a meritocracy ideology to successfully gain legitimation for their power, wealth, and privilege. For, if a meritocracy ideology becomes accepted by those in the middle or bottom of the hierarchies, then they, those in the middle or in the bottom, will come to believe that they are failures, that they had their fair chance, but they didn't 'have it' – either they didn't have the talent, skill, intelligence, or will, to make it. In defining themselves as failures, they inadvertently are defining the elite managers as successes – the elite managers 'have it', they deserve their excessive wealth, power, and privilege because they have the talent, skill, intelligence, and will. The proof of this is that they made it.

Through this meritocracy ideology, the power elite would not seem as usurpers of power, wealth, and privilege, but as legitimate, or deserving of the rewards of power, wealth, and privilege because of their superior ability and workaholic service. The meritocratic legitimation of bureaucracy is disastrous for democracy because it legitimates hierarchy and elite domination, while it de-legitimates collegiality and the peer-to-peer attitude necessary for a citizen's evaluation of a democratic leadership.

Contemporary neoconservatives,[25] who have propounded the meritocracy idea over against the idea of affirmative action quotas for Blacks and Latins and others, do not realize that this meritocracy ideology could become something quite different from that which they envision. The unintended effect of such an ideology would be to legitimate the 'New Mandarins'.[26]

Further, the meritocracy ideology, like the Confucian ideology of the ancient Mandarins,[27] is flexible enough to coexist with new ideologies of pure mysticism (paralleling those of Buddhism under Confucianism), and personal ideologies of hedonism and narcissism in the private sphere of life.

A culture of 'narcissism' in search of 'God', dominated by the 'little dictators' of the bureaucratic 'power elite', is the nightmare vision of the 'Empire Analogy'. We shall present detailed programmes to prevent this.

Let us complete the analysis first.

8 Does the Empire Analogy Hold, or Are Critical Differences Emerging?

INTRODUCTION

We have been describing the ways in which the Empire Analogy does hold for modern society. Yet, if modern society seems no longer to function according to the Polis Analogy, is it really becoming like the empires of old? There are dramatic differences between modern technocratic-bureaucratic society and the ancient empires. The analysis of these differences will surely be as crucial for an understanding of the near future world as the similarities. There are, for instance, powerful and possibly permanent sources of delegitimation and dereification operating in modern society. And, the contemporary rationally oriented population is not at all like the traditionally bound populations of the ancient empires. Rational education, rational science, and rational law may be necessary in modern society, and they may have a more positive effect on the populace than the 'iron-cage theorists' anticipated. The mass media, too, may produce multiple and divergent effects which could on the one hand help to create a manufactured charismatic aura around leadership strata which could reify power beyond our reach, while the media could, on the other hand, create the kind of cynicism and sceptical evaluation that delegitimates anything it touches.

Furthermore, modern society, while dangling the miracle of technology and the stability of bureaucracy before our eyes, may fall short of the expectations it engenders, and, as Habermas[1] suggests, be wracked by a permanent legitimation crisis wherein leaders and leadership strata would be routinely rejected rather than deified.

Before we descend into the depths of paralysing pessimism, we ought to look very closely at the actual social patterns as they are presently emerging. For the iron cage of serfdom has not closed around us, and there seem to be as many trends towards a democratic-oriented rejection of all forms of modern despotism as there are trends towards the submissive acceptance of them.

178   *The New Middle Class and Democracy*

SOME SOURCES OF DELEGITIMATION IN THE MODERN WORLD: PERFORMANCE EVALUATION OF THE POLITICAL AND ECONOMIC SYSTEM

One of the powerful analogies between the ancient empires and modern societies is the 'mega-machine'[2] ability of these societies to accomplish great and complex tasks. However, contemporary, educated, rationally oriented individuals tend not to be awed by such accomplishments. In fact, they have grown rather used to them, and tend instead to be critically evaluative.

Rather than standing awestricken at the base of a great pyramid, the modern individual compares the latest skyscraper to an earlier one he or she liked better, points out its every fault, questions whether it should have been built, and evaluates its architectural style. In the same way, the modern individual does not stand awestricken at the productive capacities of modern agriculture or machine production, but tends instead to focus on the inadequacies of the system. The overproduction of food and manufactured goods, which would have seemed miraculous to the ancients, is viewed mundanely and critically by the moderns.

What does this mean in terms of the legitimation of contemporary political systems? As Habermas has pointed out, it creates a crisis of legitimation instead of the over-legitimation of traditional empires. The crisis of legitimation occurs because our standards for performance evaluation of the system are so high that no system can live up to them. Political systems cannot create perfect order even when they become massively repressive (massive repression itself is viewed as problematic by most populations). Economic systems cannot produce to perfect capacity and with full, equitable distribution. Overproduction and scarcity will both plague economic systems, while the definition of equitable distribution itself is controversial, diverging between value orientations which favour pure communism[3] to those which favour proportional equality[4] (greater reward for greater excellence of varying kinds), to those which favour competitive inequality.[5]

Therefore, since the modern society of the future will not be a Utopia, 'systems inadequacies' will be a continuing source of legitimation crisis. This situation is quite different from that of the ancient empires.

PERFORMANCE EVALUATION OF BUREAUCRATIC ORGANIZATIONS

Another important source of delegitimation is the negative, critical view of bureaucrats and bureaucratic organizations that the general population shares in all contemporary societies.

Whereas the ancient peasants were very deferential to the ancient bureaucrats and viewed these bureaucrats as within the charismatic aura of the divine kings (and as well-educated and cultured, compared to themselves), contemporary individuals (outside, perhaps, of Japan) view the modern bureaucrats as non-charismatic, as bland functionaries, possessing some expertise and administrative skill, but possessing no general superiority over them. Therefore, the modern individual tends to dwell on: (a) the inefficiencies and policy errors of the bureaucrats, and (b) the nasty, 'officious', dictatorial, amoral attitude which bureaucrats often display towards hierarchical underlings, clients and the public.

Bureaucracy is not viewed by the modern individual as a 'megamachine', but as an inefficient, clumsy, degrading necessity of mass society. Neither Russian nor American nor Scandinavian nor Mexican moderns evaluate bureaucracy in a positive way.

If this kind of performance evaluation continues – and it may, because of certain inherent structural defects endemic to bureaucratic organizations,[6] and because of the permanent rationalization of the public worldview – then a kind of permanent delegitimation from the system may also continue. This would not militate towards a serf-like subservient attitude in modern populations, even if the moderns are absorbed into bureaucratic organizations in their work and in the service world.

ONE EFFECT OF THE MASS MEDIA ON LEGITIMATION

The effect of the mass media on the political processes of society is growing. We cannot engage in a full discussion of this remarkable new set of institutions. Here we wish to emphasize only certain aspects of the media's effect on legitimation.

In the capitalist legal-democratic societies the mass media tend to be remarkably critical in regard to the functioning of the political and economic system. The media – especially television news and newspapers – tend to focus on problems and crises and scandals: little ones, big ones, significant ones, irrelevant ones – it doesn't matter and the media do not differentiate between them.

The result is the fostering of a generalized feeling of discontent, with distrust of, and disgust for, the system and the people running it – whether they be politicians, corporate executives, or government bureaucrats. The overall effect of the mass media is to heighten the legitimation crisis, creating greater delegitimation and cynicism than might otherwise exist.

In the communist societies, the media had adopted the opposite role. Instead of pointing out every flaw, every problem, every scandal, the media presented only a very positive picture of events, processes, and leaders. But the view presented by the communist media was too positive. The people were rational and well educated enough to know that the media's view was not at all accurate, therefore they developed a sceptical attitude in reaction to the media's haloed presentation of reality (as compared to their own experience of it). In the communist countries, the overly-positive presentation of the system was simply too dissonant with the directly experienced reality. Therefore, delegitimation from the system through rational performance evaluation occurred anyway.

PARTICIPATION AND LIMITATION IN MASS TECHNOCRATIC-BUREAUCRATIC SOCIETY

Barring utopia, and given the inherent deficiencies of bureaucratic organizations, a negative performance evaluation of the managerial leadership of modern society may continue. In such a situation, the demand for more and better participatory input by the population at large may increase. Segments of the new middle class are defined by the system itself as technically skilled experts, or intellectuals, while all of the new white-collar classes are very well educated (by historical-comparative standards). Therefore, the expectation of rational participatory input into policy-making and decision-making may increase.

The key here is whether the ideology of expertise will overwhelm the idea of general education and rationality. Weber believed that the reliance on bureaucratic expertise would overwhelm the democratic reliance on 'majority wisdom'. However, if the experts continue to disagree among themselves, as they certainly seem to be doing now (for instance, on their theories of nuclear safety, inflation, the disturbance of the ozone layer in the atmosphere, genetic engineering, food additives, etc.), then it could become possible that the general population, or segments of it, may feel fully competent in terms of policy-making and decision-making as compared to the experts.

Thus, it may be that a well-educated, rationally-oriented new middle class population may see itself as capable of constructive input into the policy-making and decision-making processes of the polity, the economy and social life. In such a case, demands for the retention of, or development of, democratic participatory mechanisms will emerge from the new classes and strata of technocratic-bureaucratic society.

Such a situation would differ drastically from the Empire Analogy, and, within a new technocratic-bureaucratic social structure, would resemble the Polis Analogy in this regard.

In the same way, following the same line of reasoning, if the modern populations viewed their leadership, not as infallible experts, but as clumsy bureaucratic managers who often make critical policy mistakes and disagree among themselves, they might demand the limitation of tenure and power and, more rational succession procedures for their selection. Or, at least insist on a system of democratic electoral procedures, in order to establish a non-expert leadership system which could oversee the expert one. Elected boards of directors[7] or other 'watchdog'[8] institutions could emerge under such circumstances, along with the elected parliaments.

CLASS CONSCIOUSNESS IN MODERN SOCIETY

The 'white-collar pyramid', in all the variant forms of contemporary society, exhibits not only distinctions of authority within the hierarchy, but marked differentials of wealth and power and status as well. These latter produce the typical class distinctions which could lead to the kind of class-consciousness and class conflict which typified commercial societies.

Two processes may prove crucial in this regard:

(a) Following the Marxian formulation, whether classes develop class consciousness or not is dependent upon the ruling classes' ability to create an ideology which the subject classes accept. If the subject classes accept an ideology which justifies differentials of power, status, and wealth, then they do not develop the kind of class consciousness which would motivate them to organize or revolt against the domination of the ruling class, but instead exhibit a 'false consciousness' which makes them docile.

In the case of modern society, the rationality of the population may lead the new middle classes and the new lower classes to identify, and resent, the wide distinctions of power, status, and wealth which have been emerging as part of the new system. It is not capitalist or feudal wealth, power and status that I am referring to, but that emanating from the new system and linked to the celebrities, stars, top managers and VIPs, etc. of the technocratic, bureaucratic and media worlds.

Because of the rationality of modern populations it may be that class consciousness could emerge, such that the kind of differentials of power, status and wealth that are beginning to typify mass technocratic-bureaucratic society may be rejected, and that the kind of legally constrained electoral (and union) class conflicts that typified commercial-industrial societies could continue to exhibit themselves in the West and emerge in the post-communist countries as well. For instance, one already hears grumbling about the salaries of sports stars – who make millions now. This has led in turn to an analysis of corporate executives and their growing number of 'perks'. Comparisons between the relatively fair incomes of Japanese executives and the overblown incomes of American CEOs is being bandied about in the 1990s USA.

(b) In terms of the development of class consciousness and legally constrained class conflict, the emergence of white-collar unionism becomes important. Mills was ahead of his time with his analysis of white-collar unionism,[9] which has emerged in all the technocratic-industrial societies (and the desire to unionize existed strongly in Russia and Eastern Europe).

These unions, like most unions, are usually not democratic in their organized structure, and this does not sit well with the educated white-collar strata they represent; however, they are accepted as a necessary device for the establishment of countervailing power against the elite managers atop the bureaucratic hierarchies.

If, however, white-collar unionism becomes institutionalized, the class character of the white-collar strata could become permanently emphasized. In this case, the interplay of classes could resemble the mediated-conflict situation of the industrial-capitalist societies, rather than the status-rigidity of the ancient empires. If, however, the white-collar strata are coopted through excellent salaries and benefits, then unionism may decline.

PARTY POLITICS ON A NEW MIDDLE CLASS BASE

Guenther Roth has pointed out that Weber's and Michels' predictions on the fate of political parties in mass democracies have not proven accurate.[10] Michel's analysis of the Social Democratic Party in Germany[11] was accurate in its day; however, the political parties today have become less bureaucratic, less 'oligarchic', less dominated by cliques of professional politicians.

With the rise of a university-educated middle class, there has been a concomitant rise in party participation by amateur citizens.

In the United States, both the Democratic and Republican parties have shown dramatic alterations towards increased participation. The reform movement within the Democratic Party in the 1960s, headed by Herbert Lehman and Eleanor Roosevelt, brought the middle class permanently into the party, pushed out the old-line political bosses, defeated the city machines, and reconstituted the party along democratic-participatory lines. Then, the

Black movement was accommodated by the reformed Democratic Party. Blacks, Latins, and other poverty groups were brought into active participation within the Democratic Party. Finally, the women's movement had its impact on the party, and with significant rule changes, women were permanently brought into the party structure.

Within the Republican Party – in reaction to the 1960s and the social changes, symbolized by the Democratic Party – the old middle class asserted itself. The Goldwater-to-Reagan conservative movement drew in the Protestant fundamentalists, and the old and new middle classes of the rapidly expanding sun-belt, along with the middle classes of Texas and the South. Large segments of the conservative Catholic working class, especially the Italians, were drawn into the new conservative movement within the Republican Party as well.

Now, with the emergence of the technocratic, bureaucratic, and service middle class, the 'yuppies' – young up-and-coming urban professionals – are joining both parties, sometimes moving back and forth between them, depending on the candidate in question.

In any case, the American political parties of today are no longer dominated by Big Business and Big Labour respectively, as they once were, but are relatively open organizations whose candidate choice and platform choices are very much influenced by the participating membership. This is not to say that Big Business – the old rich and new corporate managers – are not remarkably powerful in influencing electoral choice, nor that the mass media are not crucial as well, but only that the parties, as political parties, have become open to real citizens' input. The fact that Michels' and Webers' predictions were wrong on this could provide us with a clue to the potential for democracy in mass society.

If this evidence of increased party participation were limited only to the United States, it would be less significant. However, the same process is occurring in Britain and Europe.

In Britain, although the Social-Democratic–Liberal Alliance parties did not succeed, the expanding middle class is becoming quite active. Especially the young 'red-brick' university-educated middle class seems active in today's politics. The Conservative Party is also gaining its share of the expanded middle class. Its 1990 convention was heavily participated in and more open than ever. Even the Labour Party, for years dominated by big unions and ultra-leftist political 'bosses,' now is relatively open. Since the Gaitskell years, new factions, ranging ideologically from the left to the liberal centre, have emerged with support from a more active membership.

In general, participation within the political parties in Britain has increased dramatically, and the new (1996) 'Clintonization' of the Labour Party may increase participation by the new middle class even more.

In Germany, the emergence of the Greens, 1960s-style counterculture party modelled after the American counterculture movement, has generated a great deal of political participation both within itself and within the other parties (in opposition to the Greens). The Greens created great turmoil within the Social Democratic Party, opening it up to debate, scrutiny, soul-searching, and increased participation. As with the Goldwater–Reagan Republican reaction to the liberalized Democratic Party, the Christian Democrats in Germany underwent a revival wherein the more conservative citizens flocked to their banner in reaction to the Greens and the Social Democrats. Though personal relationships within the two major parties remain more 'formal', deferential, and hierarchical in Germany than in the United States or Britain, within the German cultural context the increase in participation and the decrease in clique control has been quite significant, and, the mere existence of the Greens as an anti-authoritarian 'hippy' peace-oriented movement gives a new style to German politics which it never had – for as Michels had shown, even the Socialist parties had been authoritarian prior to the Second World War.

These same trends are occurring within France, Spain, Portugal, Italy and Greece. The Eurocommunist parties are very much new middle-class parties, open to political participation. The Socialist parties are becoming more open, while the Christian Democratic parties too are being joined by those opposing the leftist parties. In Greece, the participation level within the various parties since the decline of the Junta has been inspiring, while within Spain and Portugal the same kind of processes have emerged. Even in Japan, the recent victory of the Socialist Party has generated the need to revamp that party along new middle-class lines. In the near future, the Socialist Party could become a true opposition party, acting as a vehicle of democratic participation, to counterbalance the power of the state-bureaucracy and its LDP 'puppet' party.

Party participation in the post-communist societies of Eastern Europe and Russia has also been exciting, even though the economic instability adds the 'Weimar analogy' fear.

Thus, the dual trends continue in the modern world. The spectres of bureaucratic despotism and technocratic-totalitarianism hang over us – the 'Empire Analogy' manifests itself in modern form, casts its shadow over modern society, while at the same time glimmers of a new democratic potential fire our imagination – the 'Polis Analogy' lightening the shadow of the 'iron cage'.

SOURCES OF DEREIFICATION IN THE MODERN WORLD

By reification, as we have established, we mean the view of the power structure as a 'thing' beyond human control. The leaders, the institutions, the ideas, appear superhuman, awesome, divine.

While there are powerful new institutions in modern society which are creating the possibilities for a new form of reified power structure – such as the mass media, the computer and communications networks, and the giant size of the 'globalized' bureaucratic organizations themselves, there are sources built into mass technocratic-bureaucratic society which militate against such a reified perception of the world and the extreme alienation resulting from it. Let us introduce them now.

Science, the Decline of Religion, and Disenchantment

The Weberian analysis involving the rise of the scientific world-view and the decline of magic, ritual, and prayer in religion is now well known.[12] Weber had a strong commitment to science, but did not herald the decline of religion in the way that Marx did.[13] Weber feared that the decline of religion would leave the modern individual 'disenchanted' with the world (as well as rationally oriented to it).

The disenchantment would produce negative effects, such as existential despair and anxiety, along with a decline in the acceptance of religious ethics. The modern individual would become cynical, self-centred and self-indulgent.[14]

In this context, however, there may be some positive aspects to this disenchantment. That is, the scientific rationality, plus the cynicism in the world-view, lead also to a lack of enchantment with political leaders. The modern individual is hardly 'awed' by his or her leaders. This lack of awe means that the modern individual – even if alienated – would not view the leadership as if it were reified. The government, the leaders in any sphere of society, would not appear superhuman, they would not appear as a living fairy tale, as divine, or even as merely great.

Thus, the disenchantment with the world would involve a 'disenchant-ment' with the political leadership, who would no longer appear as magical, divine and superhuman. Weber bemoans this fact in terms of its potential effect on charismatic leadership. For, such individuals could no longer become 'charismatized', and perhaps this will have negative consequences.[15] But in the sense that such individuals may not perceive the state or the leaders of society in reified terms, so much the better. A disenchanted population is already a profoundly different population from that of a traditional empire.

We have then the paradoxical situation of the 'enchantment' created by the world of the mass media, and, the disenchantment created by the world of science and rationality. We shall have to evaluate the effect of this social-psychological paradox on the various political structures that may emerge in modern societies.

Disenchantment and Charisma (Pure and Manufactured)

The modern individual is as cynical, sceptical, critical and rationally removed from attempts at stage-managed charisma, as from pure charisma. In communist countries, the cult of the personality with its billboards, radio and television build-ups, had been greeted with increasing cynicism by the new middle class emerging therein. In the West, the Madison Avenue and TV image-makers have also had as many failures as successes in attempting to manufacture the charisma of candidates. Media celebrities have failed electorally as often as they have succeeded.

The effect of the mass media is very great, and I do not wish to minimize it. Further, more sophisticated attempts may prove successful in establishing the manufactured charisma of political leaders (and of the state as an institution). Now, at least, the disenchantment of the population with its leaders extends to the manufactured charismatic attempts, such that contemporary populations have not been won over to this kind of irrational, emotional bonding with its leadership. Contemporary populations, at this time in history, are not viewing their leaders as superhuman fantasy figures, although their feeling of alienation, or inability to control these leaders, is increasing.

Media-Manufactured Charisma and the Reified World of the 'Stars'

We have already suggested that reification of power implies that the individuals within society perceive power and leadership as beyond them – as existing in a world of superhumans in a realm to which they have no access. In the ancient society, it was the world of the kingly-aristocratic court – the sacred, secret city to which the average individual had no access. In modern society, it is the world of the mass media – especially television and the movies – which produces a separated realm of stars and superstars and celebrities, who live in a magical world separated from the world of the average individual. Once the media superstars and celebrities are viewed in this way, the reification of power has occurred.

Television, movies, magazines and radio, create this reifying effect through (1) the manufacture of charisma where genuine charisma is absent, and (2) through the enhancement of genuine charisma (where it already exists).

Although traditional forms of manufactured charisma have probably disappeared forever, leadership elites will continue to exploit new techniques for manufacturing charisma both to achieve societal integration and to legitimize their usurpation of power. We are acutely conscious today of the role of the mass media in manufacturing charisma.[16]

Leadership elites have employed specially trained social science teams and media experts to 'package' the new media-manufactured charismatic leader in the most successful way. Each of the media produces its own form of charismatic aura around the individuals who emerge as its celebrities or wish to emerge as one of its stars.

For instance, newspapers, magazines, and printed posters create an atmosphere in which a political leader seems ever-present and larger than life. Since the charismatic relationship functions best when the group feels a personal, trusting, infantilizing bond with the leader, the constant presence – in bright images – helps manufacture such leader-led relationships.

Radio is also a medium through which very successful charisma manufacture can and has occurred. Great and passionate oratory is one of the key ways in which charisma can be generated by a leader. The intersubjective communication of oratorical language can create a spell in which humans can become swayed and charismatically linked to a leader. Radio amplifies this oratorical linkage.

Hitler, Roosevelt and Churchill used radio to raise oratorical charisma to new media-manufactured heights. However, all of these leaders were also genuinely charismatic leaders who used radio to enhance their charisma. The fact that these leaders became omnipresent – that their oratory reached every ear – raised their genuine charisma to a reified realm – they were giants in the sky beyond the reach of ordinary mortals, yet touching every individual in a deeply personal way through the magic of the media.

Movies are spectacular in their ability to create manufactured charisma. Just as microphones create an oratory greater than life, so movies create physical beauty and strength far greater than life. Technicolor splendour transforms men and women into gods and goddesses possessing sensuality and strength of charismatic proportions. The movies are larger than life, the voices are smooth and beautiful, and further, the darkness of the theatre puts the individual into a dream-state in which fantasy projections and iden-tifications become easy to attain. Movie stars have an awesome potential as political figures, because they are superstars and they know how to project media-manufactured charisma (or whom to hire to help project it). We shall discuss the success and failure of such media stars shortly.

Television is still in its infancy, but is developing a potential for manu-factured charismatic manipulations more far-reaching than any of the other media. Furthermore, improvements in television relating to size, colour, clarity and cable – along with the expansion of news coverage, are com-bining to make it the most all-pervasive of the mass media in terms of its political effect.

The political power of television is already amazing. Television, through its newscasts (and newscasters) can decharismatize genuine charismatic leaders, enhance the charisma of others, create manufactured charisma where pure charisma is lacking, and create its own media-manufactured charismatic figures – stars and anchor persons – who then become national celebrities in their own right.

A test of a political leader's charisma in the modern world is whether his or her charisma can be projected through television. The kind of cha-rismatic techniques that worked well in the era of public speaking, micro-phones and radio do not usually translate well in a television appearance. Hot oratory looks out of place – it becomes annoying, grating – the gesticu-lations seem too flamboyant, facial expressions too twisted and forced, and the voice too small to justify such bodily histrionics.

Therefore, the 'great orator' does not do well on television. In fact, such oratory is rarely shown directly. Instead, with the speaker in the back-ground and the speaker's voice tuned out, the television commentator summarizes the main points of the speech with only a few sentences of the actual speech allowed to go on the air directly.

Not only is oratorical charisma neutralized, but the newsperson's inter-pretation of the main points of the speech is all that is conveyed. In the United States, only the President, and less often, party chiefs, get uninter-rupted speech time on the media and, even then, the anchorpersons engage in 'instant analysis' and critical summations of the address.

Leaders, used to 'the stump', but unskilled in the media, often look furrowed, pockmarked, and uncertain on television. Very often, they look smaller than life, as opposed to the media stars who appear larger than life. The lack of media skill, along with a clinging to older oratorical styles can decharismatize a candidate who cannot adapt to the era of television. Television, in this sense, can be a decharismatizing medium.

Some candidates, of course, have just the right combination of good looks, oratorical restraint, wit, and style that works well on the new medium, television. John Kennedy's looks and style were so well suited to television that he became charismatized through it. And, Ronald Reagan used the media as effectively as any political figure could. The projection of pure charisma through the medium is not impossible, then, but demands a different set of techniques from those that enhanced charisma prior to television.

Few political leaders have been able to imitate Kennedy's and Reagan's successes. There is good reason for this, having to do with television coverage of political leaders, rather than the leaders' failure to imitate Kennedy. Certain further decharismatizing processes have become typical in the television medium. It is now evident that even if a political leader learns to master television technique, if the television networks wish to decharismatize such a candidate, they very often can. The television newspeople tend to be cynical and often attempt to debunk a candidate. They can be formidable opponents in this regard, because they manipulate the media situation.

They may set out on a careful campaign of slandering the leader's programmes, impugning the choice of aids and friends and judgements, and uncovering embarrassing incidents from his or her personal life, as they have incessantly with Clinton. They may ask embarrassing questions during interviews, put the leader on the defensive, or never ask the questions which would allow the leader to present his or her point of view in the way they wished to.

It is because of these problems that modern political leaders have had to learn how to utilize the media to their advantage. They have had to: (a) learn how to present themselves on television when in spontaneous or arranged situations, and (b) they have had to hire public relations 'image-makers' and Madison Avenue 'marketing experts' to create television commercials wherein their image, their message, and the television situation can be manipulated on their behalf.

Madison Avenue image-making and political commercials are now commonplace phenomena. Media experts can make or break a candidate. 'Packaged' versions of candidates are presented to the public regularly. All of these image-making processes represent attempts at media-manufactured charisma. The real candidate is hidden behind the facade as surely as the kings were hidden behind their crowns, robes, and sceptres.

This is a 'rationalized' world, of course, and, therefore, the candidate's actions and policies can be analysed beyond their manufactured images.

Thus, sometimes the image-making works, and sometimes it does not. But, in either case, the presentation of the leader in today's media-dominated world is often a packaged product.

Television, then, sometimes inhibits the projection of pure charisma, and sometimes debunks it when it does emerge, while at the same time structurally encouraging the manufacture of charisma through media techniques.

The Reified World of Stars and Celebrities

If television debunks the pure charisma of political leaders, it also produces charismatic figures of its own. That is, television, like the movies, produces stars. Television stars are, of course, packaged personalities. We do not know much about the 'real' person. The star possesses some modicum of charm, beauty or other characteristic of importance in that not everyone who is packaged catches on. But the greater part of the personality is manufactured image.

Stars have become potentially powerful as political leaders. C. Wright Mills, so far ahead of his time, wrote of the celebrities, and how easily they mingled with the politically powerful figures of the United States. He was so right when he suggested that the 'power elite' would include figures from the world of the celebrities.

Figures like Ronald Reagan and others, have already emerged as political leaders. This process has begun to occur in Italy, France and Germany as well.

Media celebrities, whether they be movie stars, television stars, sports figures-like Jack Kemp or Bill Bradley, or astronauts like John Glen are taken seriously as political leaders even though they have no political experience. They are taken seriously because they possess media-manufactured charisma. This manufactured charisma produces an irrational bond between such a leader and a mass audience in typical charismatic fashion. This irrational bond can be utilized to create mass acceptance in the political sphere. This does not mean that any media celebrity will necessarily become a successful political leader. Many such celebrities have failed, and will continue to fail, in their bids for power. John Glen did not win the presidential primary, neither Bradley nor Kemp became president. Robert Redford might lose a senatorial race in Colorado or Utah if he ran. The initial factor is not whether such media celebrities win or lose, but rather that they are taken seriously by the public as candidates for political office, and they take themselves seriously as such. Furthermore, the ability of a celebrity to become an 'instant candidate', even though they may have little previous political experience, is certainly a new phenomenon to be reckoned with.

As media celebrities emerged as political leaders, television anchor people themselves began to stand out as stars – and, therefore, potential political leaders. As television news ratings skyrocketed and news shows expanded from one half-hour to two hours and from morning to evening to night-time (with slots in between), and now with whole cable networks devoted to news – CNN, CNBC, for instance – the anchor persons themselves have been manufactured into stars. Their faces are everywhere, they command prime-time and they 'chaperone' every critical event of the moment. Not only do they enhance or debunk the charisma of political leaders, but they can become political leaders themselves. If they do, they will do so as television celebrities.

Millionaires, if they have a celebrity family name, can become instant television celebrities. Nelson Rockefeller, Winthrop Rockefeller, Jay Rocke-feller, any Kennedy, have used their family name – coupled with the mass buying of media spots – to create instant political stardom for themselves. The use of personal wealth to purchase media time and media experts adds another dimension to the reification of power. Ross Perot is a perfect example here, as is Steve Forbes.

Again, such attempts by millionaires may fail, but the fact that they become instant candidates and can enter the reified realm of the 'stars' is a dangerous phenomenon for democracy.

The age of television has just begun. Size, colour, and cable, if newly combined, could produce a 'hot' medium of the future in which the manu-facture of charisma could reach heady heights again.[17]

Television technology, on the other hand, could, instead of leading toward '1984' media-manufactured charismatic political leadership, lead toward a new kind of rational legitimacy process in which, through local and national cable TV hookups, a new kind of participation could be generated in which media-monitored town meetings and debates could create a new kind of democratic politics.

KNOWLEDGE OF HISTORY, DEMOCRACY AND LAW

A potentially important difference between modern populations and those of ancient empires is the knowledge that modern individuals possess of history, democracy and law. The individuals of ancient empires knew nothing of democracy or non-religious law, nor did they conceive of the world in historical terms (that is, in terms of change).

It is possible that these ideas and institutions could be inconceivable again within a soulless world of bureaucratic-technocratic efficiency, order and abundance.

For instance, it is possible that education in the modern world may become purely technocratic-administrative and non-humanistic (such as the Frankfurt School fears). In such a context, democracy, law and humanistic morality could become viewed as antiquated, or at least irrelevant under modern conditions. Huxley, in *Brave New World*, believed this could happen. But this latter is, after all, negative utopianism, is it not?

Negative utopianism or not, the 'Empire Analogy' seems more likely to emerge than the 'Polis Analogy' under modern conditions. Somehow, it is easier to envision a soulless bureaucratic despotism[18] – or worse, a technological totalitarism *à la 1984*[19] – than a new form of democracy which would survive bureaucratic rationality and somehow utilize the new technology to facilitate democratic participation rather than subvert it.

However, since bureaucratic authoritarianism and technological totalitarianism are not yet upon us, let us examine the possibilities for democracy in a mass technocratic-bureaucratic world. At this moment in history, we still possess the knowledge of democracy and law, do we not? So, let us look at the prospects for democracy in this high-technology global village.

But first, let us examine the particularistic effects of specific political cultures on the democratic process. Then we can return to our generalized analysis of the technocratic global village and mass-mediated world culture.

9 Political Culture Against Democracy

INTRODUCTION

In this treatise, I have emphasized class balance and its link to democratic stability, focusing especially on Aristotle's theory of the middle-class majority and its supportive effect on democracy. I have not emphasized political culture. Yet, obviously, the political culture of a nation is a causal factor in creating democratic stability or inhibiting it. In fact, the culture of a nation may be so non-consonant with democracy as to prevent the adopting of democracy, even when the structural conditions of that nation are ripe for the institutionalization of democracy.

The most famous cases of this were those of Germany, Spain and Italy in the 1920s, and Argentina in the 1950s and 1960s. Now, the question of the potential for Asian democracy is being debated. Is the culture of China too non-consonant to support democracy? Is the 'Confucian ethic' inhibitory to democratic processes of debate, criticism and competitive elections? Is Japanese democracy only procedural, or is it real? Will Asian democracy – if it emerges – be different from European democracy because of Asia's dramatically different cultural and civilizational characteristics?

Along these same lines, the question of Islamic religious and political culture has been debated. Are there inherent ideological and structural tendencies emergent from the Muslim religion which actively inhibit democracy? The whole problematic of religious ideology and its political impact is opened up once again. The old debate over the link between protestantism, capitalism and democracy, and the Catholic inhibition of democracy, has been revived in these new debates and has served as a parallel to the analysis of Islam and Confucianism. Since these questions are being actively debated in the West in the 1990s, let us look more closely at the concept of political culture, as it relates to democracy.

POLITICAL CULTURE AGAINST DEMOCRACY

Lucian Pye,[1] writing on Asia, has recently revived the concept of political culture, and its consonance with, or inhibition of legal democracy. As Samuel Huntington puts it, it has been argued that the world's great historical cultural traditions vary significantly in the extent to which their attitudes, values, beliefs, and related behavior patterns are conducive to the development of democracy. A profoundly anti-democratic culture would impede the spread of democratic norms in the society, deny legitimacy to democratic institutions, and thus greatly complicate if not prevent the emergence and effective functioning of those institutions.[2]

According to Huntington, 'the cultural thesis' comes in two forms. The more restrictive version, articulated by George Kennan, states that only 'Western' culture provides a suitable base for the development of democratic institutions. A less restrictive version of the cultural obstacle argument holds that only certain non-Western cultures are peculiarly hostile to democracy. The two cultural nexes most often cited in this regard are those influenced by Confucianism and Islam.[3] The reader may also know that Catholic culture was, for a long time, believed to be a staunchly anti-democratic force, such that nations such as Italy, Spain, the Philippines, and the Latin American countries were held back from democracy by the Catholic religious orientation to the world.[4]

There is no doubt that certain cultural milieus are more hospitable to democratic process than others. However, the varying cultures, their consonance or non-consonance with democracy, and the possibility for change within cultures, must be carefully analysed.

CLASSICAL GREEK THEORISTS

At the outset of this discussion, we wish to establish that neither ancient Greek nor Enlightenment theorists relied on the concept of political culture. Greek theorists such as Thucydides, Plato, Aristotle, Xenophon and later Polybius, were debating the best form of government for the Greek city-states. A unified Greek culture was accepted as a given for all Greek city-states – save perhaps for Sparta, which was acknowledged to have developed a more restrictive, more purely military, more archaic Greek pattern.

Plato, in describing his ideal Republic[5] uses no cultural causality, but rather focuses on the creation of a well-educated 'guardian' class – chosen for their intellectual capabilities. And, this intelligence factor was locked in by casting down the children of the guardians who did not measure up intellectually, and by raising up those children of the non-guardians who did show intellectual prowess. No racial, ethnic, or class bias is mentioned by Plato. In fact, startling for his era, he shows no gender bias either – brilliant girls were to be raised up to guardian status, while dull boys were cast down.[6]

The question of whether resident foreigners should be allowed political participation was raised by all the Greek theorists, and, in these debates it becomes clear that such foreigners were recognized as capable, and that it was the question of their loyalty that inhibited the Greek theorists from extending citizenship to them.

Aristotle did show a Greek chauvinistic bias when he asserted that the middle Eastern peoples were too slavish for democracy – tending to obey their traditional kingships without questioning this authority, while the Europeans, though spirited and independent-minded like the Greeks, were too illiterate and primitive to adopt any of the various forms of Greek-style self-governing forms of polity.[7] Nonetheless, Aristotle recognized the Phoenicians to be both learned enough and independent minded enough to establish a rational-democratic form of government. His account of the Carthaginians constitution – which we have but a fragment of in *The Politics* – does not make any cultural distinctions important, though the Carthaginians were semitic-speaking Phoenicians, and, of course, non-Greek. The Phoenicians were well-educated in letters and numbers, having, of course, brought the alphabet to the Greeks. And, the Phoenicians were independent businessmen and sea merchants, living in polis-style city-states.

Further, in his chapter on education in *The Politics*, Aristotle makes it clear that, given the leisure time necessary for a proper education, the rational-mindedness and independence of thought necessary for democratic (or mixed oligarchic-democratic) polities could be instilled in a populace, living in a polis-sized political unit.[8]

THE ENLIGHTENMENT THEORISTS

The ancient Greeks were debating about the best form of government for the city-states, with little attention focused on the non-Greeks. The Enlightenment theorists in Scotland, England, France and the United States, however, seemed to be arguing that any nation could establish 'the rule of law', and a representative parliamentary system. From their works one gets the impression that the overthrow of the monarchy would *ipso facto* lead to the establishment of a legal-democratic state, and that such a state would be stable. No mention was made of the cultural – or for that matter the class-structural – factors that might undergird such a change. Such

Enlightenment theorists as Hobbes, Locke, Hume, Mill, Montesquieu, Rousseau, Jefferson, Madison, Hamilton, Paine and others focused on the political structure alone. This was the case with Machiavelli as well, whose works were influential with the Enlightenment theorists.[9] The class structural elements were, of course, tangentially debated within the context of the oligarchic versus democratic system of representation for the parliament, and with the question of the feudal aristocracies' relevance, or irrelevance, to the new constitutional arrangements. But cultural elements, as causal factors influencing forms of government, were not made central at all.

The Enlightenment theorists did not believe that French, German, Italian, or whatever culture, was a barrier to the institutionalization of democracy. Neither the English, nor the Scots, nor the French or American theorists dwelt on cultural differences, but rather on universal, inalienable rights, derived from nature and God.[10] The rights of man[11] – natural rights – God-given rights – were the centrepiece of their theory.

A nation had to establish the constitutional-legal processes – elections, representative bodies, courts, separations of power, limitation of power and terms of office – that insured that legal-democratic procedures would be established and maintained.[12] Further, a nation had to provide a proper education for its citizenry, and then guarantee the freedom of speech, press, and assembly necessary for the citizenry to carry out its democratic political functions.

As with Rousseau,[13] and Jefferson, many, though not all, Enlightenment thinkers believed in the educability of all men, and, that an educated citizenry was the proper foundation for a democratic polity. No theory of culture intrudes here – an educated citizen is an ideal citizen, whether English, French, Dutch or German.

THE CULTURAL ARGUMENT BEGINS TO INTRUDE

The eighteenth-century theorists were fully aware of the difficult and violent transition to democracy that had occurred in many of the Greek city states – they had read Thucydides.[14] They believed that a period of violent revolution against the monarchies and the feudal lords in Europe was immanent. But the long-term result was clear to them: the legal parliamentary state, in one form or another, would eventually triumph. No notion of 'political culture' or 'national character' or religious inhibition intruded on their political prediction.

Why should the eighteenth-century theorists have dealt with the concept of political culture? Weren't British and French cultures different and, yet, were not both nations monarchies? As a matter of fact, the English 'puritanized' middle and upper classes found French culture to be shockingly sexual and frighteningly emotional and uninhibited. Still, both England and France seemed to be moving similarly toward the same political goal: the revolutionary clash between aristocracy and monarchy on the one side, and the common people and legal representative government on the other. How would any theory of 'political culture' intrude into their philosophical framework?

Later, however, once the French Revolution engendered so much terrible violence and fratricide (and as a side issue, sexual excesses, such as those graphically described by the Marquis de Sade), notions of national character and political culture began to be thought about. Even then, however, British theorists, such as Edmund Burke,[15] preferred to fall back on purely political theory – theory embodying the contrast between slow orderly evolutionary transitions, contrasted with sudden revolutionary breaks with the past. For Burke, for instance, the American War of Independence was viewed as successful in its outcome because it represented a moderate transition beyond the English Civil War. The French, on the other hand, were attempting too great a leap – from true monarchy to complete democracy all at once. Thus, anarchy, chaos and violence resulted.

The cases of Germany and Spain, in their transition from feudalism and monarchy to legal democracy, began to bring out the causal significance of 'political culture' beyond that of the British and French cases. For the German and Spanish intellectuals, along with the average person on the street, held a great distrust of democracy and a burning attachment to the glory of the aristocratic warrior (the 'knight in shining armour') and his king.

Spain and Germany are extreme cases in which the culture of the nation was extraordinarily nonconsonant with the processes of a democratic political system. In fact, so nonconsonant were these cultures with democracy, that their leading theorists either opposed and belittled democracy, or insisted that democracy could never be adapted to the cultural conditions of the nation.[16] It must be remembered that democracy, when first established in Spain and Germany failed, and that the dictatorships were more popular with the majority of the people than the democracies had been. Even the greatest humanistic theorists of these nations, such as Miguel de Unamuno[17] in Spain and Max Weber[18] in Germany, doubted the possibility for establishing and stabilizing democracy in their respective nations. Unamuno spoke in the newly established Parliament, shocking his supporters by admitting that he did not think that parliamentary procedure fitted with the Spanish cultural character. Weber admitted to Michels in a letter that he did not think that 'Anglo-Saxon conventions' could be adapted to Germany.

Furthermore, democracy had to be forced on Germany by the victorious Allies after the Second World War. Nor was there a major resistance movement against Hitler and the Nazis in Germany. Of course, Spain surprised the world in its peaceful and orderly transition to democracy after Franco's death. But this transition, made so easily when it finally occurred, makes two points we wish to establish.

One, the political culture and structure surrounding Spain and encompassing Spain's immediate 'civilizational nexus' had changed dramatically. That is, rather than being attached to a divided Western Europe, half-democratic, half-fascist – with the fascists in the ascendant and the English and French democrats in a period of decline following the First World War – Spain was now surrounded by a remarkably modernized Western Europe. Western Europe had become democratic and capitalist and was now economically prosperous and politically stable. Furthermore, American influence had become profound. As the victor over fascism, the United States became the most powerful nation in the world, and served as a model for economic and political emulation. Since such emulation had produced very positive results in Germany and Italy, Spain looked to reproduce those positive results. The Spaniards quietly joined the Common Market (later the European Union) and established joint ventures with Italy, France and Germany. Spanish fascism (watered down under Franco), had opposed democracy as weak and disorderly, but had never opposed capitalism. Therefore, the expansion of capitalist enterprises was allowed and was later encouraged by the Franco regime.

Two, the socioeconomic structure of Spain had changed dramatically. From a semi-feudal agrarian economy, Spain evolved toward a modern capitalist economy in the cities and the countryside. By the 1980s, Spain began to look strikingly like Northern Italy and France in its economic structure and life-style. Spain was alive with cars, trucks and motorcycles, television sets and stylish clothing, and all the other consumer items typifying Western Europe. Many of these products were made in Spain, with joint-venture partners from the European Union. The success of the capitalist economy engendered a rapid expansion of the middle class, both white-collar and entrepreneurial – which became well-educated and prosperous. They role-modelled their EU counterparts quite closely. The old aristocracy was still around, the poor peasants still existed, the 'aristocratic' style of Spanish culture and its focus on daring and death still remained, symbolized in the bullfight. Yet Spain was changing. One did not have to be an expert to see the change. Tourists visiting Spain in the

1980s were amazed and delighted by the modernity, lack of political repression, and resurgence of Spanish artistic and literary life on a non-fascist base.

The transition of Spain, in terms of both indices (that of the altered cultural nexus surrounding Spain, and the socioeconomic developmental process within) is highly significant, for it serves as a potential model for China, the Pacific Rim, and Eastern Europe. Two great differences exist though – not in the cultural, but in the structural realm. That is, the post-fascist nations already contained well-developed capitalist economies, and had post-feudal parliamentary institutions as well – this latter in the *standestaat* era, or era of 'estates', with the third estate the democratically elected burghers of the free-trade capitalist cities.[19] Spain had a capitalist economy, though, of course, it was not very extensive. And, Spain's Cortes,[20] or national parliament, was one of the first parliaments in Europe, existing full blown in the sixteenth century.

In Germany, the Hansa towns had pioneered trade capitalism in Europe, bringing it from Italy into the essentially feudal north. These Hansa towns also brought plebeian democracy – guild democracy – to Europe from Italy (wherein such democracy had been overridden by wealthy oligarchic patricians). Thus, even fascist Spain and Germany contained capitalist and democratic institutions largely absent in China and Russia (though existent to a lesser extent in Czechoslovakia and Hungary). The lack of any significant capitalist or parliamentary institutional structure in the communist and Asian nations has made the transition to democracy much more difficult.

Yet, who could be foolish enough to assert that the transition to legal democracy in Spain and Germany was an easy one? Let the Graeb memorandum[21] on the German concentration camps, and the descriptions of the slaughter of one village against another during the Spanish Civil War[22] remind the reader that the transition to democracy in China – Tienanmen Square massacre or not – might be smoother than that of Germany and Spain, though the political culture and the social structure may have been less consonant at the beginning of the process.

Still, cultural and structural factors are important. The case of Japan, for instance, reminds us that democracy – set in a samurai–Mandarin bureaucratic structure, with a Shinto–Confucian cultural orientation – may be greatly restricted in terms of the electoral participation and civic debate.

CONSONANCE AND NON-CONSONANCE

Some political cultures are more consonant with legal-representative democracy than others. Where there is consonance, the transition will be easier.

German authors, such as Weber, made much of the fact that English culture, known for its politeness in debate and cool reasonableness in controversies, was perfectly adapted to parliamentarism, whereas the German culture of domination and military action, was not. Weber, as mentioned, wrote of 'Anglo-Saxon conventions' conducive to democracy. As we have seen, however, Germany, Spain and, to a lesser extent, France, with cultures non-consonant with parliamentarism, did eventually adapt to it successfully. The non-consonant cultural elements certainly do make the transition to democracy more difficult.

Within our framework of analysis, however, it should be kept clearly in mind that the political culture is carried by certain classes who reflect certain specific world views such that an alteration in the class structure toward the commercial and salaried middle classes[23] (with their education to the rational-scientific world-view) can and will change the political culture of the nation in question.

France, for instance, has become more bourgeois in its upper class and less aristocratic. Its middle class has expanded, become better educated and more business and professionally oriented. Some tourists in the 1990s are complaining that France is not as interesting anymore – it is no longer the nation of love and *haute cuisine*. In fact, the American-style shopping mall now has replaced one of the markets once full of great food, love liaisons and prostitutes. We have already mentioned the dramatic class and cultural changes in Spain and, such changes have been reported in Hong Kong, Singapore, Korea, Mexico and Brazil, and other rapidly capitalizing and democratizing nations of the world.

Still, the presence of the non-consonant classes and their political culture makes the transition more difficult.

In Japan, for instance, the transition to legal democracy occurred in form, but not in function, for a very long time. Some scholars, as we have shown, believe that Japan still does not have democratic procedures operating at any deep level.[24] Yet, the recent fall of the majority party shows that the parliamentary system is developing in a democratic direction. And, more positively, the use of the law courts by citizens to defend and protect their individual rights has been increasing. Furthermore, opposition political parties – as they once defended the rights of workers and small farmers (the socialists and ruling party respectively) – may turn toward defending the rights of the expanding middle class.[25]

China, like Japan, may retain non-consonant cultural inhibitions to democracy. These non-consonant factors were generated by the Confucian–Mandarin traditions, and, in modern form, by the communist system.

Note carefully that both the Confucian–Mandarin as well as communist traditions in China have deeply ingrained structural as well as ideological components. Therefore altering the Chinese political system in a democratic direction will involve both a structural and ideological change of great magnitude.

With the movement towards capitalist industrialization, however, the business classes and the educated professional strata may override these cultural tendencies.[26] Already, for instance, contract law and a stock market have sprung up, and the Chinese middle classes are avidly absorbing rational-scientific educational programmes, overseas and at home.

NATIVIST MOVEMENTS

If political culture will not help predict long-term trends, it certainly can tell us what kind of 'nativist' movements may emerge during periods of transition, and what modifications and peculiarities in the democratic process of government may emerge.

We have already shown how the knightly-military traditions of Germany and Spain led toward fascism as a nativist movement (though, of course, it first emerged out of Italy). The rise of Muslim fundamentalism in the Arab countries and Iran is another example. The success of the Ayatollah in Iran is a perfect example of a nativist movement with a special cultural link. American officials and Russian officials misread the Iranian situation, because neither set of officials could conceive of a Muslim-oriented regime in a modernizing state.

The Iranians would probably be equally bewildered by American fundamentalist Protestants – the radical religious right – in the United States. Yet, this movement is electing local officials and may elect governors in various states – and, of course, Reaganism was, in large part, American Protestant 'nativism' in its political–cultural content.

In any case, nativist movements – like the Mau Mau in East Africa, the Shining Path in Peru, anarchism in Italy or Democracy in Greece – will be specific to that nation's particular and peculiar political culture. Therefore, though world movements, such as communism, capitalism, or fascism, will sweep into a nation, peculiar nativist movements will also emerge, and in some cases they will lead to long-term modifications of the political process.

In the ancient world, for instance, when kingship swept through the Middle East, the ancient Jews and Greeks resisted this mainline trend, engendering in the first case a religious–purist pastoral rejection of kingship (see the Book of Judges and Prophets), and in the second case engendering the oligarchic councils and democratic assemblies of Ionia or the council of elders of the Doric city states.

With ethnic particularism flaring up all over the world, nativist movements will probably also emerge. The Slavic areas, the Caucasus area, Africa, and so on may engender specific nativist movements, such as we have never before encountered.

One last note: charismatic leaders[27] are often linked to nativist movements, for, such leaders, using the nativist cultural mores, strike a 'sacred' chord in their followers – usually a chord which outsiders simply cannot hear, because they are out of harmony with it. So, the Germans screamed and cried when Hitler spoke, surrounded by his regimented, uniformed guards, while British, French, and American journalists simply thought Hitler was a madman and his brownshirts should be arrested. And the American State Department officials never took the Ayatollah seriously, hearing what they thought was the blitherings from some past theocracy of a tenth-century Caliphate.

Charismatic leaders can, and do, lead modernist movements, but, often enough they emerge as nativist leaders – or, at least, surround themselves with the trappings of nativist culture (which can be anything from a leopard-skin headdress in Africa to Ghandi's Hindu cloak in India).

RELIGION, CULTURE AND DEMOCRACY

Along with culture in general, the specific effect of particular religions on political organization has been the focus of much discussion and debate in the modern era.

If we look at the continuum of theoretical usages on cultural influences on political organization, we note that three categories of culture are currently in use: (1) a civilizational nexus, that is, broad area categories, such as Asian culture, Hispanic culture, Western culture, North West European culture; (2) national culture, that is, German, French, English, American, Japanese culture, and so on and finally (3) religious cultural influence, that is, Catholicism and Protestantism, Confucianism and Islam, etc.

PROTESTANTISM, CAPITALISM AND DEMOCRACY

The whole field of scholarship focusing on the link between religion and emerging socio-economic systems was generated by Max Weber's brilliant, yet controversial, thesis on the link between puritan Protestantism and the origins of capitalism.[28]

Weber's argument, as he states in his author's introduction, was not meant to be a refutation of Marx's thesis on material causality for social change, but rather an attempt to create a balanced view of the relationship between material and ideological influences on social change.

In terms of puritan Protestantism and its link with capitalism and democracy, Weber focused on these elements: Protestant individualism in both the interpretations of scripture and in the personal relationship with Jesus and God. This individualism was unimpeded by, and antithetical to, any church hierarchy, and even any ministerial control of the local congregation. Such an individualism was consonant with individual business ownership and economic competition and development, without any interference from the state bureaucracies.

This unfettered economic individualism and individual interpretation of scripture was also fully consonant with the independent-minded citizenship demanded in the emerging democracy movements.

Now, whether Protestantism helped 'cause' industrial capitalism and legal democracy, or whether they simply emerged with it, or were caused by it, is for the Weberians, Marxists, and others to continue arguing about. But that Protestantism is consonant with, and therefore supportive of, modern capitalism and democracy – this is not debatable.

Before we go too far with this consonance, however, let us look at Lutheranism.

LUTHERANISM AS ANTI-DEMOCRATIC

Protestantism split. Wherever Calvinism prospered, industrial capitalism prospered. And, where industrial capitalism predominated, legal democracy emerged triumphant.

The Lutheran churches emphasize the 'work ethic' – hard work in one's calling can gain one heavenly grace; sloth is of the Devil. But Lutheranism did not emphasize money accumulation as a sign of grace, nor did it encourage the puritanical denial of sensual pleasure insisted upon by the Calvinists. Business and puritanical wealth accumulation were not emphasized. Work and military-feudal obligations were blended into a less progressive ethic that supported the feudal-monarchical state, and linked the work ethic to political obedience and military heroism.

Therefore, Lutheran Protestantism did not generate direct pressure towards democratic political processes. Industrialism, literacy and hard work

were emphasized. And scientific and mathematical thinking were generated. Along these lines industrial technology was also accepted. But all of
these latter were located within a framework of state power and citizens'
obedience to feudal–military authority.

From the case of Lutheranism, one can see that religious ideologies can
generate change, but they also can be adapted to differing social conditions, supportive of differing socio-political systems.

CATHOLICISM AND DEMOCRACY

Seymour Martin Lipset[29] and others have analysed the anti-democratic
elements of Catholicism. What I would like to emphasize, however, is the
structural, as well as ideological system embraced by Catholicism, which
tended toward anti-democratic political processes.

It is important to understand that Catholicism emerged from the fall of
Rome, and as supportive of the feudal-kingly political system in medieval
Europe.

If Catholicism was wedded to feudalism, it was also at odds with it.
For, the Catholic church had its own hierarchy, headed by the Pope. Thus,
there emerged the 'Cesaro-papist' separation.[30] The Cesaro-papist separation and the conflicts it engendered – such as the Guelf–Ghibelline
struggles that racked Renaissance Italy – did create a system in which the
separation of powers became understood, if not accepted.

However, though the separation of powers would become a hallmark of
later Enlightenment political doctrine – already prefigured in Machiavelli[31]
– it must be understood that the Catholic church supported kingship and
feudalism, and would actually oppose popular democracy, secular law and
the money economy that would arise with trade capitalism. The Church also
opposed the scientific and rational-philosophical thinking of the Renaissance, and the individualism and free thinking of the Reformation.

The Catholic church did not change its position until after the Second
World War. Even then, in the surveys conducted,[32] Catholic individuals
exhibited less of a democratic inclination than the Protestant citizens of
the same nation. In France, Italy, Spain, Austria and Bavaria, the Catholic
church opposed democracy from the sixteenth to the twentieth centuries.
Even the 'liberation theology' of the 1970s and 1980s in Latin America
was more socialistic and humanistic than it was democratic.[33] Liberation
theology is a wonderful throwback to the Jesuits' attempt to save the
Indians back at the time of Columbus.

Finally, though the present pope has helped use the power of the Catholic church to bring down communism in Eastern Europe, this same pope has attempted forcefully – charismatically – to control the thinking of Catholic citizens, in the democracies, on a whole range of social issues.

Thus, Catholicism remains in an enigmatic relationship with modern democracy, and may still be non-consonant with democratic ideals.

ISLAM AS AN INHIBITOR TO DEMOCRACY

At the outset it must be established that within many of the Islamic nations the majority of the population tend to be poor and uneducated. Therefore, we need not look to Islam as the basic cause of the anti-democratic trends in those societies.

However, the contrast between India and Pakistan, for instance, provides an example where Islam does seem to exert an anti-democratic pressure, as opposed to Hinduism, wherein the Brahman ethics were more consonant with democracy ('Untouchables' aside, of course).

And well-educated intellectuals within the Islamic nations often exude strongly anti-democratic sentiments and remain in search of 'Islamic solutions' to modern political questions.[34]

So, let us look more closely at the ethic of Islam in order to isolate some of its anti-democratic elements.

The Koran emphasizes, over and over again, submission to God and the judgement day. The pious Muslim must submit to Allah, get down on his knees, follow the laws set down by Mohammed, and not question the good set down in the Koran. Question the laws, waffle in your faith, involve yourself in worldly excesses, and you shall burn in hell.[35]

This Islamic ethic hardly encourages the kind of independent thinking required of a democratic citizen. Catholicism, until recently, espoused a similar doctrine, discouraging free thinking, 'indexing' taboo books and ideas. Protestantism, as has been pointed out, encouraged individual interpretation of the scriptures and tolerated sectarian differences.

Islam, to its credit, however, has always encouraged wise commentary on the Koran and the Bible, and has remained relatively open to various philosophical interpretations of scripture. For instance, the works of Aristotle were preserved and interpreted by Muslim theologians – eventually influencing Thomas Aquinas – even though his rational argumentation and metaphysics were not Koranic or biblical.[36]

Still, Aristotle's logic, science and philosophy were considered as 'Greek thinking', and even though Muslim intellectuals made great breakthroughs in mathematics and science, these were carefully separated from Islamic thought.[37] The split between 'Greek science' and Islamic morality brings us to another anti-democratic element in Islam. That is, the centuries-long hostility between the Muslim world and the 'Western' world.[38] One of the longest wars in history has been that between those who converted to Islam and those who retained their Christianity or Judaism.

Any new reader of the Koran will be surprised to discover that an on-going, repetitive dialogue occurs between Muhammad and the Jews and Christians. Muhammad first recognizes the wisdom and holiness of the Old Testament and the New, accepting Moses and Jesus as Apostles of God. But, then he demands that the Jews and Christians accept him as the new apostle of God and convert to Islam.[39]

Since neither the Jews nor the Christians of Europe and Spain converted, a holy war was sanctioned by the Muslim wise men. From before 700 to the present, the war has raged. From the time of the crusades to the Gulf War, Islam has been on the defensive. To modern Moslems the war with the 'West' is still very real – both politically and ideologically.

Now, not only are Judaism and Christianity the targets of Islamic zeal, but democracy and capitalism are targets as well. So deeply ingrained is this anti-Westernism, that many Muslim intellectuals continue to actively seek Muslim alternatives to democracy and capitalism.[40]

The Ba'ath Party movement embodied these ideas, as have the more extreme groups who are still committed to the holy war, or *jihad*. And, of course, the American domination of the oil industry, and the immigration of the Jews into Palestine have fuelled the anti-Western *jihad* all across the Middle East.

The Middle East, however, is modernizing, and as it does so, the relationship between Islam and the modern economy and polity will probably improve. Islam has always been an intellectual religion. The Islamic wise men have historically been able to study science and metaphysics while combining them with Koranic teachings.[41] If anything, over the centuries, Islamic toleration has exceeded that of the Catholic church. Therefore, if the war with the 'West' cools down, the Muslim religion may emerge as compatible with capitalism and democracy. Finally, an expanded, well-educated middle class could go far towards engendering the kind of accommodation between Islam and the secularized modern society that has occurred in the Catholic countries – even while the search for a modernized Islam continues.[42]

A Note on Women and Islam

Though traditional Islam has women veiled and locked into their households, it is important to note that modern women – once they were liberated enough to receive a college education – were well accepted within the modernized segments of Islamic nations. Muslim women have become journalists, academics, successful business entrepreneurs, and, even political leaders.

The easy acceptance and success of Muslim women within modern Islamic society, and their frightening repression within fundamentalist societies, such as Iran, points up both the potential for an Islamic success within the modern context, and the traditional barriers to that success.

CONFUCIANISM AND THE BUREAUCRATIC STATE

Confucianism represents both a code of ethics for individual conduct, and a political ideology supportive of the mandarin-style kingly-bureaucratic state. Confucius, writing during the sixth century BC, created a beautiful ethical system based on the key concepts of benevolence, reverence for parental authority, and acceptance of feudal kingly-bureaucratic authority.[43]

In terms of personal ethics, the concept of benevolence supports a caring and empathetic attitude towards the needs of others, and a deep concern about the vicissitudes of life and the need for people to support others. Confucian benevolence is a paternalistic benevolence, very much framed within class, age, family, and gender political roles. It is a caring and humanistic ethic, nonetheless.

Like Platonic philosophy, Confucianism involves one in a life-long quest for wisdom and knowledge. And, as with Platonic philosophy, Confucian works are presented as dialogues with the master. And, just as Plato was focused on the quest for 'justice' and political order among men,[44] Confucius was focused on 'caring' within the political order.

The two great thinkers, however, part ways once we come to their analyses of the socio-political system in which the justice and caring are acted out. For, Plato – reacting negatively to the revolutionary activities and civil war that were tearing Greece apart in the fifth and fourth centuries BC – questioned every basic principle upon which political association was built, and theorized in their stead a utopian republic of his own creation. The questioning, perhaps more than the utopianized solution, is the basis of the greatness of Plato's work.

Confucius, on the other hand, though confronted with a good deal of feudal disorder – which directly affected his life, since he was attached to the feudal court of the Duke of Chou – did not question the foundation of human political associations. Rather, he attempted to create an ethical system which would allow the feudal-kingly-mandarin-bureaucratic system to function at its best: that is, benevolently.[45]

If the officials were caring and humane, if they were not self-serving and greedy, and if the local feudal lord and the emperor worked for the good of the people – in terms of *corvée* projects which were economically beneficial, and military preparedness – then the system itself would confer benevolence, and therefore be ethically acceptable. If, however, the officials, the local lords, or the emperor did not act benevolently, then the people had to right to remove them.

This latter is a remarkable point of view for any thinker analysing this kind of political system, for traditional authority systems never allow for any interventions by the people, tending to dominate them through 'divine right'.[46] However, as remarkable as Confucius' idea of removal of a self-serving or incompetent aristocratic or bureaucratic leader may be, it is not to be confused with revolutionary thinking or action.

For no change in the political, economic or social system is demanded by Confucius. Rather, he demands that a new emperor, local lord or government official be appointed who will have the people's needs foremost in their minds, and who will therefore govern in a benevolent manner.

Thus, though Confucianism provided a powerful corrective to monarchical and bureaucratic excesses – a corrective which did assert itself from time to time in Chinese history – Confucianism does not question the social order. And, in fact, the Confucian ethic of benevolent paternalistic rule became the legitimating ideology for the Chinese system.[47] So powerful was the wedding of Confucian ethics with the Chinese political system, that it was not questioned in any basic manner, from the time of Confucius until the days of Sun Yat-sen.[48] And, even then, the more consonant system of communism – with its host of government officials and promise of benevolence – became more popular, more understandable, than the individualistic, seemingly selfish, limited-power-state of the proposed capitalist-democratic alternative for China.

ANTI-DEMOCRATIC ELEMENTS IN CONFUCIANISM

Rule by Educated Bureaucrats

Confucius recognized and accepted the distinction between the gentlemen-officials, who ran the state, and the common people, who laboured in the fields, workshops, and *corvée* projects.[49] He also fully accepted the military aristocracy, who ruled regionally and existed in quasi-feudal vassalage to the emperor. Confucianism was thus not only an ethical system, but also an ideology supportive of a particular kind of political system.

No conception of citizenship for the common people was allowed within the Confucian system. The common people were conceived of as hard-working and basically good, but they lacked education – the many years of profound education – that it took to create a gentleman-official. Since the people lacked education, the well-educated, literate officials would rule for them, under the mantle of the emperor's authority.[50]

The notion of rule by an educated elite is hardly unique to Chinese society. This was Plato's ideal – his guardians – and the ideal of the British gentry who believed that their oligarchic rule was better than a democratic alternative in which the uneducated common people would have an equal vote. However, the Confucian ideal of rule by the well-educated literati officials is set within the structure of the monarchical state – a centralized bureaucratic state, with no conceptions of lawful protections of the common people or limits on the power of the emperor and his vassals. This was not Plato's *Republic* or John Stuart Mill's *Representative Government* (with extra votes for the well-educated).[51] The Confucian state was the kingly-bureaucratic state. The educated officials were hierarchically organized in a chain of command that culminated in the centralized elite power of the emperor's court.

The shadow of Confucius still falls across modern Japan and Singapore, as well-educated bureaucratic officials dominate the polity, even while representative institutions exist in form, and in partial function.

POLITICAL HARMONY AND FILIAL PIETY

Confucianism emphasizes social harmony: harmony between classes, within families and between individuals. It actively strives to reduce conflict. All political ideologies seek to reduce civil and revolutionary conflicts. However, those raised with Confucian ideals find even the legally-constrained conflicts within parliamentary democracy to be unharmonious.

Political factions, parties, debates, and egoistic candidates – all these seem disruptive and unethical to the good Confucian. Lee Kwankue[52] of Singapore, for instance, becomes self-righteous and morally offended when presented with the notion of a rival political party or citizen's action association. Singapore, as he fashioned it, has a benevolent state committed to economic development, a rise in the living standards for all, and political order. What more could one want? Why disrupt it with contentious debates, irresponsible speech and press criticisms, and the criminal disorders that emanate from a less authoritarian state?

Whenever Lee is criticized by American and British scholars, Lee shoots back with Confucian pieties centred on harmony and benevolence.

The leaders of China's Communist Party hit back with the same

Confucian-oriented answers to the students demanding legal-representative democracy at Tienamen square. There would have been disorder, they insisted, if they had given in to such demands.[53]

Japan is still deeply rooted in a combined Confucian–samurai ethic; when the Japanese finally elected a prime minister from the opposition socialist party, the situation so unnerved the parliamentary politicians that the new prime minister created a united cabinet, and, suggested disbanding the socialist party entirely and creating a new party more closely unified with the government party (and more middle-class-based than working-class-based). Harmony, rather than conflict were emphasized immediately. This was deeply Confucian in orientation.[54]

With the Confucian ethical system, then, parliamentary debating is considered impolite, and extremes of political position are frowned upon. Remember, however, that this traditional Confucian attitude can be modified with modern educational exposure and with the inculcation of legal authority. Weber, you will recall in twentieth-century Germany, as well as Unamuno in Spain believed that the Germans (and Spaniards) could not engage in parliamentary debate because they were likely to challenge each other to a duel, rather than argue politely as the British do. Still, the Confucian rejection of political conflict, even within legal-parliamentary confines, remains powerful throughout Asia.

What of the age and family factors? In the Confucian ethic, sons should obey fathers, young men should attach themselves to older men, who become their mentors and sponsors. The elders should be listened to, deferred to, where they have wisdom – and the wisdom of experience-in-age was accepted as a natural phenomenon.

One can see then, that a system wherein the old would lead the young would be more acceptable than a system where the voting age makes everyone equal in wisdom.

Look at China's leaders now! And, even in Japan, where there is forced retirement from Japan's bureaucracy, the retirees do not retire, but rather become either parliamentary leaders or business leaders, as I have described in an earlier chapter.

WOMEN AND CONFUCIANISM

The exclusion of women from both family and political power and authority is also extraordinary in Chinese and Japanese culture, even for traditional societies in general. Little or no mention of women occurs in the work of Confucius – they seem not to exist. The bound feet, concubinage, and female infanticide practised by the Chinese set their women at a lower status than most traditional cultures. It should not be assumed that all traditional cultures were equal in this regard.

Herodotus[55] reports that Egyptian women were often politically powerful and very directly domineering in sexual matters as well. The Roman writers, such as Tacitus, in *The Annals*,[56] makes a point of declaring that Roman women were held in higher regard than the Greek women.

The Jewish Bible, though denigrating women in many of its ethical-mythic tales,[57] makes women heroes of countless episodes in Jewish historical life, with Miriam, Debra, Judith, and Esther powerhouses of Jewish ethical goodness. And Eve, Bathsheba, Delilah, and Jezebel, domineering seducers of men and demonic destroyers of Jewish ethical purity.[58]

Thus, the complete dismissal of women within the Confucian ethic should be seen as an extreme of male domination and female submission, even among traditional societies. Even Hindu women – though forced to die on their husband's funeral pyre – if they were upper-caste Brahmins, acquired high status and great respect. This was absent within Confucian cultures, the wives and daughters of the gentlemen remaining as far in the background of political and social affairs as those of the common people.

However, it must be mentioned again that the liberation of women in all male-dominant societies has been a slow and painful process, and therefore the Confucian ethic should not be seen as an impossible barrier in this regard.

BENEVOLENCE, SOCIALISM AND DEMOCRACY

The ethic of benevolence is the central ideal in Confucianism. Politically, it is interpreted as assuring that the common people have their basic needs fulfilled, and that the lords and officials set aside their personal ambitions and rule for the good of all.

Given Confucianism's political emphasis on benevolence and deference, one should not be surprised to find that communism had a deep appeal in China, and that the Japanese, though understanding the greater efficiency, productivity and creativity of capitalism, did set limits on the wealth distribution process, and established a special system of benevolent caring for its workers and farmers (though during the Second World War, the cruel, quasi-fascist regime brutalized the Japanese population as the war effort became more desperate).

The point here is that the Confucian ethic of benevolence – which is beautiful – does inhibit notions of rugged individualism, citizen-assertive political action, and selfish wealth accumulation. The Japanese, however, have created a fascinating amalgam of Confucian benevolence and corporate-capitalist productivity. This amalgam has been dramatically successful economically, and the question remains, will full-fledged democracy emerge from this amalgam?

MERITOCRACY VS DEMOCRACY

The Confucian ethic arose as part of a system of state bureaucracy in the traditional Chinese kingly-feudal political system. Central to it was the meritocratic system of examinations through which officials were chosen. Technically speaking, the examinations were open to all young men, no matter what their class background.[59]

In practice, the sons of the officials and aristocrats were better prepared for the examinations than the sons of the peasants or artisans, and therefore gained easier access to these high-status positions of authority. However, many bright young peasant boys studied hard, and did gain entrance to officialdom. Therefore, the ideal of fairness was maintained. Thus, this meritocratic system of administrative bureaucracy was accepted by the Chinese. And the officials were seen as having earned their positions of authority through study and intelligence.

The common people, in accepting Confucianism, lent their consent to the officials. Confucianism legitimated the mandarin, meritocratic-bureaucratic state of China very effectively. And, of course, Confucius himself was a local administrative official.

Now to our point: this Confucius-legitimated meritocratic system divided the society into two markedly different strata. The one, the common people: passive, deferential, uneducated and uninvolved in any governing role. The other, the gentleman-literati officials, in complete control of the political and economic processes (in all except military affairs).

This is significant for us here, because this system – reinforced strongly by the Confucian ethos – is anti-democratic. For, with the Confucian meritocratic system, the everyday people became separated from the political process. They remain subjects, and cannot become citizens.

Notice that the consent of the people can be 'given' even in a system where the people, in so consenting, give away most of their political rights.[60]

What is the key factor here? The key factor is that the Confucian ethic is linked with a structural institution – the mandarin-style system of state, with government officials selected through meritocratic examinations.

Where this system is fully adapted to the modern high-technology capitalist economic system – as in Japan – it regains its legitimacy because it helps to successfully and benevolently administer the capitalist economic system.[61]

This, of course, is supposed to be an impossible and contradictory phenomenon. Adam Smith would have nay-said it in 1776, and both Milton Friedman and Mikhail Gorbachev would have agreed in 1976. The failure of the communist-style bureaucracy to administer a modern industrial economy seemed to prove the *laissez-faire* thesis correct.

However, the Japanese, and now the Singapore Chinese, have created a new pattern – they have linked a Confucian-style meritocratic benevolent bureaucracy with a capitalist-industrial economy. All of the dynamics of capitalism are institutionalized and followed. The bureaucracy, as we have described it, 'guides' the capitalist system, where it shows instability or where it needs special support.[62]

The success of this Asian model in economic productivity could inhibit fully fledged legal democracy from emerging in Korea, Singapore, Thailand, and China.

Even though modern rational-scientific education has replaced Confucian-style education, the examination system in Japan has created the same kind of two-tiered system as in the Confucian–mandarin societies of the past. An elite group of bureaucrats administers the polity and economy, while the 'common people' remain uninvolved.

This is the potential Asian model of modern high-tech economics and politics, which could become the model for much of Asia, if China adopts it.

Given what China is like now politically, the adoption of the Japanese model would seem quite wonderful! And, of course, Japan and other Asian nations may see their political systems evolve toward more typically democratic patterns, as modern education sweeps broader and deeper into the new generations, and as the mass media intervene more fully in the socialization process of the young (this media process, of course, containing as many negative, as positive, influences on youth).

It is possible, though, that even with education and the world mass-media culture, the modernized version of Confucian meritocracy could continue to exert a structural influence on the Asian state and Asian citizenry, as it does in Japan.

Look carefully here. In modern society there are two trends emerging simultaneously: one, the majority of the populations are becoming well educated. Such an educated population will feel qualified for political participation,[63] and will feel like an Aristotelian citizenry whose majority wisdom is often more astute than that of the expert few.

On the other hand, there is a growing tendency among the educated common people to withdraw from politics and leave it to the experts. This latter trend could lead to a Confucian–mandarin drift in all modern societies. Even in the USA, where anti-government feelings are at an all-time high, and where bureaucrats are reviled, fewer citizens are voting and fewer participating in political organizations than ever before.

Now, to take an optimistic tack, we cannot jump to the assumption that Japan will be the Asian model. As Fukayama[64] has asserted in a recent article, Chinese Confucianism has never been as submissive to centralized authority or as hierarchical as the Japanese samurai adaptation became. And, furthermore, according to Fukayama, the modern rational-minded university education is inculcating in the Asian intellectuals that same feeling of qualification for political participation that swept across Europe after the Renaissance.[65]

Thus, we should take seriously the dynamic democratizing trends in Taiwan,[66] Korea,[67] Hong Kong and Tienanmen-China. Furthermore, Singapore is a new nation. Its founding leader is still alive and its population is barely 25 years into modernity. The democratic movement in Asia could become quite militant, as at Tienanmen Square, or among the Korean university students.

Let us conclude this chapter on political culture, then, with an analysis of the sources of change, within our emerging high-technology global village.

CULTURAL CHANGE

Civilizational Nexes and Political Change

Political culture is a national phenomenon, but it can also be part of a civilizational pattern. That is, one can speak of Muslim nations, or East Asian culture, or Western culture, or Hispanic culture. There will be national variations within any cultural nexus – such as French vs British, or Chinese vs Japanese – but the cultural configuration in its broad sense can still be helpful in understanding political trends.

For instance, the Asian deference pattern (especially in terms of age) makes parliamentary debate and elections seem rude and inappropriate. Hispanics, deep down, want a machismo leader, personal and strong – one who overrides law and social rules.[68] The Muslim cultural pattern would exclude women from politics and public life in general.

Sometimes, therefore, a cultural nexus will exist and this system of values and norms may inhibit or encourage democratic development in the entire region.

We have also suggested, however, that when there is a basic change in the cultural nexus – as say, with Asia going capitalist or Europe going democratic after the Second World War – this change can draw in all nations of that cultural nexus toward that change. In this regard, we have seen Spain go democratic, and we may witness China go capitalist and then democratic, as the nations in her cultural nexus move in that direction. As the 'four tigers' (Hong Kong, Taiwan, Singapore, Korea) and Japan become increasingly successful at capitalism and democracy, China may be drawn in, as Spain was. In this sense, the eventual success of Mexico, Brazil and Argentina, and the failure of communism, Peronism, and militarism could have such a 'civilizational' effect in all of Latin America.

CLASSES AS CARRIERS OF CULTURE

One of the structural phenomena that alters political culture in the long run is the emergence of new classes carrying new institutions, norms, and values. New classes can emerge as the result of the general evolution of the 'mode of production' – as say, the high-technology industrial economy has engendered 'new' middle classes, such as service, technical and managerial classes.[69] Or, new classes can be created politically – as say under communist regimes which created the new middle classes through the establishment of a modern educational system before the industrial economy itself had been fully instituted. A new class structure can be engendered, also, by military domination, as in the case of Japan after the Second World War. The United States supported the business class in Japan, while at the same time weakening its samurai roots with a tough land reform programme. This latter not only weakened the old samurai aristocracy, but also strengthened the independent farmers, who then backed the business party, giving them an electoral advantage over the socialists.

This slight shift in the class structure of Japan altered the culture of Japan just enough toward business and away from its purely military orientation.

Along with economic, political and military factors, the class structure of a nation can change through diffusionary trends. Economic institutions, as well as political and ideological orientations, can be adopted from neighbouring societies which may be influential. Small business in Greece, for instance, is right now adopting EU standards of production and distribution.

The alteration of the small business class in Greece – from petty shop-keeping to modern franchising, has had a profound effect on the economy and polity of Greece, moving it toward democratic stability and corporate prosperity.

No matter whether the structural changes in society are brought about through economic, political, military or diffusional changes, once they are brought about, they engender new classes. And when and if a new class emerges – whatever its origin – that class carries with it a culture of its own, which then begins to permeate the entire cultural complex.

Thus, the cultural affinities of the new middle class that emerged in the communist nations because of the Soviet educational system began to permeate that culture. If a large commercial middle class emerges, say, in China, as it has in India, then this class's culture will affect the whole cultural complex.

Conversely, if classes decline or disappear, their cultural affinities decline with them. So, the military aristocracy of Europe has largely disappeared, and the European culture has become more bourgeois than it used to be, more business-oriented, less 'macho' and 'Don Juanish'.

With the decline of the industrial working class, in the high-technology nations, the culture of the working class – along with socialism, of course – has declined. The high-tech and service middle classes cannot remember why anyone would have found Marxism exciting. They can't hear the music of socialism, because the stereos in their BMW or Lexus surround them with the sound of their individual affluence (to mix metaphors).

Thus, the culture carried by specific classes becomes critical to the overall cultural context of a nation, and the cultural affinities of classes must, therefore, be understood, along with the class balance in any given nation.

The class balance of a nation is absolutely critical in terms of both the stability of the polity and the form of government that may result from it. Central here is Aristotle's formulation that a nation with a majority middle class will likely be stable, while a nation sharply divided between rich and poor will likely engender oligarchy or tyranny.

To Aristotle's theory of the middle class we have added that the character of a specific class – or culture which a class carries – is also crucial to the political system. For specific classes will carry specific cultures which influence the long-term direction of changes in a given society. So a military upper class will lean toward one type of polity, a business upper class another.

Therefore, both the culture carried by a given set of classes, and the class balance between them in that society, will influence the political culture that emerges.

With the concept of class balance, we have come full circle to our original analysis, developed here and in our earlier volumes.[70]

THE GLOBAL VILLAGE: WORLD CULTURE

Concepts like civilizational nexus or national political culture may become *passé*. The spread of the global economy, and the mass media, may create a world cultural nexus to which everyone will become socialized (in part). Of course, specific cultural differences will remain deep, but it is possible that an overriding set of institutions, norms, values, roles and beliefs may emerge with the world economy, its classes, its high-tech orientation, and, especially its mass media of communication.

The 'world system' is happening – no longer as a capitalist hegemonic power trip, but rather as a world-integrating process. Trans-national corporations are shaping the world; CNN, and other news networks, are making the world's populations aware of each other in an immediate way that has never occurred before. MTV, CDs, movies and VCRs have created a youth culture that has become an adult culture, that shares a pattern of consumerism, dress, music and recreational tastes.

In the summer of 1993, visiting the small city of Kalamata in Greece, my wife and I were astonished to find her relatives' children watching television – with the same American shows our children watch, playing the same video games – Japanese-made, wearing the same tee-shirts, jeans and baseball hats, doing the same dances, and competing in the same sports (such as basketball and beach volley-ball). Twenty years ago, we were struck by how different her relatives' culture was from our own. Now, the similarities overrode the differences. Their house was our house, their children's games our children's games, their car our car – most importantly, their emerging concerns and values our concerns and values.

It was startling – and though it was equally startling to cross into the former Yugoslavia and watch the indigenous political culture override the world culture – it was clear that the world cultural nexus will emerge, and that it will have a global effect on the next generation – for better or worse. Just exactly what the content of this new world culture will be has not yet been determined, but that consumerism and pop culture are now dominating it, is clear. That these latter are linked to capitalism is clear; that capitalism undergirds legal democracy is clear; how high-technology capitalism, orgiastic consumerism, and the pop culture of the mass media will effect legal democracy, is less clear. Positive trends, such as the desire for free expression, are visible, but negative trends, such as the media's ubiquitous use of violence, sexuality and sensationalism to hold our attention are also visible.

Finally, the United Nations, with its legal-democratic charter, is also part of the emerging world culture, and, of course, we hope it can have as powerful an effect as the mass media.

CONCLUSION

If political culture is, indeed, causally significant – and we think it is – then it follows that along with structural changes, cultural changes must be introduced that will help engender, stabilize and reinforce democracy. In this light, we have included a later chapter on 'education for democracy' in this treatise.

One last note for the moment. It is obvious to anyone living in the final decade of the twentieth century that ethnic conflict can become hideous and genocidal, and, that ethnic conflict can destroy democracy. Yugoslavia has torn itself apart, and Europe was racked with ethnic megalomania throughout the early part of this century.

Wherever such ethnic hatreds and conflicts erupt – such as in Rwanda, Burundi or Sri Lanka – demagogic-dictatorial-tyrannical rule will often emerge.

Thus, a multi-cultural world is wonderful and terrible at the same time, and, as arid as the mass media are, the global village they create may be a welcome respite from ethnic hatred.

Part IV

Participation, Power Limitation, Law and Democracy in Technocratic-Bureaucratic Society

PARTICIPATION, LIMITATION AND LAW IN MASS TECHNOCRATIC-BUREAUCRATIC SOCIETY

Overview

The thesis of this work is that the new middle class, as a majority middle class, could become the base for a new form of democracy if new democratic political institutions emerge and old ones are extended. Just as new democratic institutions emerged in the nation-state, beyond those of the city-state, so must they emerge in the mass technocratic-bureaucratic state if democracy is to be refashioned and maintained. The representative parliament of the nation-state is a very different political institution from the popular assembly of the city-state (or the council of clan-elders of the tribe).

There were those, like Rousseau, who believed that only the direct democracy of the popular assembly (such as that of his beloved Geneva) could work. He was not completely wrong. The participation level of democracy was severely impaired by the institutionalization of representative democracy. On the other hand, legal protection for the citizens and limitation upon potential despotic leadership have functioned quite well within the framework of parliamentary representation.

Before going further with our analysis of new forms of democracy, it should be mentioned that new forms of despotism emerged within the nation-state beyond those exhibited within the city-state (or the kingly empires). Therefore, one could expect that new forms of despotism will emerge in technocratic-bureaucratic society as well.

Already the outlines for the new despotism are taking shape. Technological surveillance devices and computerized data files have created a communications revolution with frightening political implications, while the authoritarian tendencies within the growing bureaucracies have increased as the scope of their activities has been extended into all but the most private areas of social action.

Realizing that new despotic institutions are likely to emerge, one can still ask the question, what kind of new democratic institutions have been necessitated by mass technocratic-bureaucratic society, and, how can the democratic institutions which still exist in the capitalist-industrial societies be adapted to the new social institutions?

10 The Extension of Legal Authority over Public and Private Bureaucracies

Government and corporate managers now control vast organizations that reach into the everyday lives of all modern citizens. Yet, these leaders are not elected, nor is their tenure limited. Of course, such bureaucrats are subject to constitutional and local law, but they have in many cases gained unique powers not regulated by the usual laws.

For instance, in the case of government bureaucrats in the United States, J. Edgar Hoover, when he was chief of the FBI, gained power beyond the usual constraints of our constitution. As is now known, he 'blackmailed' presidents, such as Kennedy and Johnson, and intimidated Martin Luther King, Jr. Yet, he was a very effective administrator of the FBI.

Now, some Americans will defend Hoover and others will attack him. But, the fact is that no clearly spelled-out limitations of his power and tenure of office existed.

To use a more mundane case, a highway bureaucrat – Robert Moses of New York City – used his position to direct and control the construction of bridges, highways, parks and recreation areas. Moses gained so much power – though he was a non-elected bureaucrat – that his vision became the reality of much of the structure of the New York area. Now, in retrospect New Yorkers often complain that his powers were too extensive and his tenure too long.

How do we prevent government bureaucrats from becoming latter-day Bismarcks? Of course Bismarck was not elected, and of course Germany was not a democracy then. But now, even with democracy, government bureaucrats often gain power and control beyond the intent of their office. Look at the Japanese bureaucrats who run MITI. They literally lead Japan. Perhaps the Japanese constitution should include legal limitations upon their power and tenure.

Given the increasing power of seemingly benign government bureaucrats, it may be necessary in modern society to specifically limit the power and tenure of all key bureaucratic offices of state.[1] Of course, in democracies, the chiefs of bureaus are often replaced when a new party comes to power. The problem with this is that not all chief bureaucrats are replaced, and party alternation does not always occur. Therefore, a regularized limitation of the tenure, and an explicit limitation of the powers, for all chief bureaucrats has become necessary in modern society. With the increasing tendency to rely on specialized experts, this limitation of power and tenure is all the more needed.

LIMITING THE POWER OF CORPORATE MANAGERS

Many corporations are transnational, others are near-monopolies – there has been a trend toward corporate mergers and giant size conglomerates. From Japan, where the giant conglomerates have long been institutionalized, to the USA, where they are vilified in principle and in law, the giant conglomerates exist and are expanding. The Europeans are rapidly creating their own huge firms to compete with their Japanese and American counterparts.

Given the size and scope of these corporate units, and the number of people they impact as employers and producers, it may be necessary to create some sort of power limiting or power-checking mechanisms through which corporate managers and financial owners are controlled.

Again, as with government bureaucrats, we are dealing with non-elected leaders. And, again, even though they are technically under the constraint of law, in actuality they make decisions affecting our environment, our employment, our technology, and our life-style that are monumental.

Theorists such as Robert Dahl[2] have suggested that we create elected boards of directors. Such boards would be elected by employees, stockholders, and local communities. Boards of Directors used to act as a check on corporate managers, because they were made up of the stock-owners. This kind of check on the power of managers still exists, but only in that the financiers who have bought up the stocks have tended to fire managers eliminated through mergers.

But, who can limit the power of these financiers? And, where the managers buy up their own companies' stock, how can they be limited, when they then become the board of directors?

Given this change, and the dual threat of financial or managerial autocracy, employers, consumers and communities have become helpless in terms of exerting any countervailing power on the new high-tech giant corporations.

So, perhaps Dahl, and other theorists such as Ralph Nader,[3] are correct. Perhaps it is time to recreate the Board of Directors as a democratic institution representing the interests of the employees and the impacted community, and acting as a check on the power of both managers and financiers.

Would a democratically elected board of trustees allow a management team to give themselves huge bonuses when profits were down or thousands of employees fired? Would a democratically elected board allow a financial team to break up an efficient, profitable, technologically advanced unit? Would pollution be better controlled?

On the other side of the argument, would such a system inhibit the entrepreneurial creativity and advancement of the free market capitalist system?[4]

Whether a democratic board of directors would be ineffective or inhibitory is not known at this time. One fact is known: corporate managers have gained control of tremendous wealth and power, and so too have the financial cliques that have been buying their stock. Neither the corporate managers nor the financial stock dealers are old fashioned entrepreneurs. Since their power and wealth have increased dramatically, it probably is prudent to think in terms of creating mechanisms to contain their power, wealth and tenure, while still allowing the market mechanisms of competition, pricing and profit to function well.

In case the reader does not understand why the Smithian 'invisible hand' is not enough of a constraint on modern corporate leaders, let us remind him or her of the change in title of top corporate leader. From businessman to corporate executive, the situation looked similar. But, from corporate executive to CEO – chief executive officer – we get a change that is a quantum change. A chief executive officer sounds like a power-leader, not a business leader. And, if business leaders have become 'chiefs', then we had better limit their power.

A NEW BILL OF RIGHTS PROTECTING INDIVIDUALS IN THEIR RELATIONS WITH BUREAUCRATIC ORGANIZATION

A new bill of rights has undoubtedly been necessitated by the emergence of a new threat to human rights – the bureaucratic organization. The new bill of rights would protect individuals in their relations with bureaucratic organizations in the same way that the old bill of rights protected individuals from government. The new bill of rights would apply to all bureaucratic organizations, private or public. It would provide protection for: (1) Those employed by bureaucratic organizations, (2) Those serviced by bureaucratic organizations, and (3) Those impacted by bureaucratic organizations, such as consumers and community dwellers.

It may be helpful to our discussion, if before describing the new bill of rights, we describe why the old bill of rights was established in the first place.

Ira Glasser, director of the American Civil Liberties Union, summarizes the philosophy behind the Bill of Rights in this way:

> The Bill of Rights was devised to protect individual rights against the excesses of well-intentioned, democratically elected political rulers. Rights were defined rather simply as limits on government power. To say that a citizen has the right to distribute a leaflet or worship freely literally meant that the government was without the legal power to stop him or her. . . . The Bill of Rights focused not upon the good intentions of democratic rulers, but rather upon the harm to individual rights that might flow from their excesses. In fact, the Bill of Rights assumed their excesses and sought to limit them. . . . These amendments say what government may not do.
>
> Those who advocated the Bill of Rights (believed that) if explicit limits on government power were not included, government officials would surely go beyond their enumerated powers. The Bill of Rights was a product of fear.[5]
>
> What was required to protect liberty were legal and political structures that, once erected, would be difficult to tear down . . . Thus the protection of individual liberty was seen as growing from an adversarial process. Government was established by the consent of the governed, but government was nonetheless an adversary of the governed, and a dangerous one at that.[6]

The expansion of government, corporate and service bureaucracies has already been described as a social fact and is hardly controversial. But, if it is true that an adversarial posture must be assumed in order to limit power and safeguard citizen's rights, then this adversarial structure must be extended directly against those new government and private institutions which have developed enormous new power potentials.

A BILL OF RIGHTS FOR CORPORATE EMPLOYEES

Like the state and municipal governments, the giant business corporation is a government. By employing thousands of individuals, it possesses the power to rule them. It can establish employment rules restricting their conduct; grant or withhold financial rewards otherwise unavailable; or effectively destroy the career of a specialist in a monopoly or near-monopoly industry. How can a Constitution, which fully restrains all levels of political government from invading the rights of citizens, then permit every business corporation to do so?[7]

Potentially the giant business corporation should be subject to the fundamental restraints that the Constitution places upon states and federal government, including the right of the corporation's citizens, its employees, to participate in the governance of their workplace.[8] Occasional protection of union free speech is not enough. There are numerous cases of corporations firing or otherwise penalizing employees who participated in political activity – on or off the job – or who wrote or published critical articles or expressed unpopular opinions.

Retaliation against outspoken employees occurs most frequently for 'blowing the whistle'. The conscientious employee risks losing his job for reporting practices such as marketing of defective vehicles to unsuspecting consumers; the waste of government funds by private contractors; the industrial dumping of mercury in waterways; the connection between companies and bribery or illegal campaign contributions; or the suppression of serious disease data. Men and women who should be lauded as public citizens have been intimidated and ostracized.[9] This has recently occurred in the tobacco industry in the USA.

In the USA, under the First, Fourth, Fifth and Ninth Amendments, the state may not invade certain rights of individual privacy. These amendments create 'zones of privacy' which government officials may penetrate only under extraordinary circumstances – such as threats to national security or 'probable cause' of the commission of a crime. Otherwise, they may not penetrate at all.[10]

Under the First Amendment, the federal government normally may not question a job applicant about his religion, political opinions, or past political association. Here, ruled the Supreme Court, 'the views of the individual are made inviolate . . . the opinions of men are not the object of civil government, nor under its jurisdiction . . .'.

The Fourth Amendment guarantees the right of the person to be secure in their persons, houses, papers, and effects, against unreasonable searches and seizures.

No such niceties apply at the workplace. Eighty per cent or more of our largest corporations subject the applicant to a battery of personal and psychological interviews and tests. As a condition of employment, the job applicant must answer inquiries respecting such non-job-related topics as his or her reading or travel habits, non-work interests, religious faith, relationships with parents, mental difficulties, homosexuality, sexual fidelity or abnormality, political views, 'loyalty', and what Sears, Roebuck once quaintly referred to as 'values'.[11]

Many corporations go much further. Former FBI agents or organizations such as Fidelifacts or Bishop's Service are hired to learn what they can about an applicant by talking to neighbours, former employees, and co-workers. A home interview may also determine whether the applicant is 'controversial': even polygraph tests may be used. On-the-job surveillance often becomes intensive as well.

Furthermore, industrial espionage has become widespread – something like a corporate CIA is becoming institutionalized, utilizing sophisticated bugging and tapping devices including wiretapping, hidden cameras, spike and parabolic microphones, remote sensing devices, and infrared photography.

Finally, though the Black movement and the Women's movement have brought great gains to these and other groups who have been discriminated against, Congress has never fully applied the concepts of equal protection and equal opportunity to the largest corporations.[12]

> True, it has long been a cherished value in this country that an owner may employ his property as he wishes. But this makes little sense when applied to supercorporations. The executives who control corporate property do not own it. . . . A giant corporation is more like a public utility than a person's home. . . . A giant corporation, unlike a human being, has no inherent rights. . . . When we say the corporation should be 'constitutionalized' – that is, the corporation should be made subject to applicable principles of the Constitution – we are asserting that . . . the legitimacy of our constitution itself is at stake. . . . The Constitution is seriously devalued if not undermined, when important activities of American citizens are not protected by its guarantees.[13]

What would an employee Bill of Rights look like? Corporations, to the same extent that the Constitution requires of the United States Government, should in every transaction, practice, or occurrence:

1. Observe the First Amendment requirements of freedom of religion, freedom of speech, freedom of the press and peaceable assembly;
2. Respect the rights of privacy of its employees and all other United States citizens;
3. Not discriminate on account of race, religion, creed, or sex.[14]

A BILL OF RIGHTS CONCERNING SOCIAL SERVICE AGENCIES

Just as Nader, Green and Seligman make a strong case for a bill of rights for corporate employees, Ira Glasser, Director of the American Civil Liberties Union, makes a strong case for a bill of rights protecting citizens from social service agencies.

The government has become much more than the political institution of the state; it now also includes the social institutions of caring: public schools, mental hospitals, public housing authorities, health insurance and health-care organizations, centres for the retarded, nursing homes for the aged, welfare agencies for the poor and more.[15]

Beyond the purely governmental social services agencies, the private social service agencies exist, in most cases, as 'quasi-public' social service agencies, as an extension of the governmental system.

Yet, we built no legal restraints into the delivery system of social services. We became oblivious, in the context of social services, to the adversarial relationship between power and liberty, and we assumed that the interests of the clients were not in conflict with the interests of social service agencies.[16]

> Vast discretionary power thus came to be vested in an army of civil servants (and quasi-civil servants), . . . organized into huge service bureaucracies, which began quietly and silently to trespass upon the private lives and rights of millions of citizens.[17]

The discretionary power faced by the aged in nursing homes, for instance, also has traditionally been faced by residents of other service institutions, such as patients in mental hospitals, or children in foster-care institutions. Similar intrusions flowing from similar discretionary powers were also to be found, though to lesser degrees, in the administration of public schools, public housing, and public welfare, and now managed health-care facilities.[18]

Because social service professionals were presumed to be acting in the 'best interests' of their clients, no one thought to question the excesses of their power. And so, a tradition grew up. The Bill of Rights existed, but it did not apply to service institutions. It limited the powers of elected representatives, political officials and the police, but it did not limit the power of high-school principals, social workers, housing officials, or mental health professionals. Those citizens who lived under the jurisdiction of the service professionals were consequently without rights.[19]

WHAT WOULD A BILL OF RIGHTS FOR SOCIAL SERVICE AGENCIES LOOK LIKE?

1. *The right to control personal property*: Unless residents of 'total institutions'[20] can exercise control over their personal property, they will be at the mercy of the institution.
2. *The right to control your own body*: The general law of torts, which basically regards as an assault any medical procedure carried out on an individual against his will, has been expanded to include the individual's right to receive all relevant information he would want to know before consenting to medical treatment.
3. *The right to come and go freely*: Here we are referring to nursing homes, but not to prisons, with mental hospitals somewhere in-between.
4. *The rights to free speech*, association, petition, and counsel must be guaranteed over against the rules and regulations of the service organization.[21]

THE NATIONAL OMBUDSMAN

The Swedish office of Ombudsman has been misunderstood in the United States and Britain.[22] The office functions as a protector of the people from the abuses of government officials, and has been extended to cover abuses stemming from the military, economic and service bureaucracies as well. In the 'common law' countries, courts and lawyers and legislators are supposed to perform this function. The problem is that they have not effectively done so because of the massive structural changes that have emerged.

Therefore, modern democracies, no matter which tradition they adhere to, should institute a national ombudsman office specially empowered to: (a) investigate, on a regular basis, the uses and abuses of power by giant bureaucracies, and (b) prosecute cases of abuse uncovered by the ombudsman, or brought to the attention of the ombudsman by an individual citizen.

The ombudsman and his associates would be chosen in the same way that Supreme Court justices are chosen in the United States (however their appointment could emanate from Congress (Parliament). The ombudsman is not above politics, but is independent of politics once appointed (as are Supreme Court justices).[23]

Further, the ombudsman must be completely independent from all bureaucratic organizations. Instituting ombudsmen within the power and authority structure of a bureaucratic organization defeats the purpose for which the ombudsman's office was created. In the United States this mistake is consistently made.[24]

Lastly, for the ombudsman's office to function properly, the courts must be attuned to administrative law precedents – which, in the USA and Britain, is an undeveloped area of legal authority.

In today's society in which government, military and corporate bureaucracies have grown enormously, the mere existence of an ombudsman, independent of the bureaucracy, to which anybody may carry his complaints, will act to sharpen the attention of the authorities dealing with cases to counteract tendencies toward abuse of powers and arbitrary decisions. In the administrative field – where there is no counterpart to the civil law statute on court procedure – there are many questions of principle which, in order to be treated by statute, have to be investigated in the light of actual practice.[25]

Modern governments, military establishments, and economic corporations – modern bureaucratic and quasi-bureaucratic administration – have grown so enormous that an expansion of the Ombudsman's activities is undoubtedly warranted. In fact, separate ombudsman's offices for civil, military, service, and corporate affairs have been established. In Sweden and Germany there has already been a call for the expansion of, and intensification of the Ombudsman's sphere of authority. It has been recommended that special ombudsmen be appointed for the tax system, for social affairs (children's care, temperance care, and social welfare), and for the municipal authorities. (There is already an ombudsman for military affairs, as mentioned.) The need for such an enlargement is beyond all doubt.[26]

It must be made clear, however, that such ombudsmen's offices would all have to be established as fully independent from the administrative systems which they would oversee. The prosecution of the citizens and the prosecution of officials could not occur as intended without such independence. Nor is such independence utopian. Though the courts, for instance, have been sometimes bought off and influenced in the common law countries, their record of independence is not so bad as the cynical predictor might have imagined back in the days of their establishment. And, especially where the middle class majority has been prosperous and well educated, the legal system has been fairly responsive to all the citizens involved. The new middle-class majority could keep the ombudsman system of people's defenders equally as independent as the courts, if the power of the elite administrators is properly limited.

11 Mass-Mediated 'Direct Democracy': Television Town Meetings and Computer Referenda – Civil Society through a Lens

PARTICIPATION IN POLITICAL DECISION-MAKING

Participation has undoubtedly been the weakest sphere of activity in the legal-democratic nation-state. For, though the representative institution allowed for the expansion of democracy from the small city-state to the large nation-state, direct democratic participation in decision-making was eliminated.

The electoral-representative institution has lent stability to nation-state democracy – stability beyond that of the popular assemblies of the polis. The reason for this is that the 'crowd psychology' and 'mob rule' political processes are greatly reduced with the introduction of the representative council. The speeches of demagogues and the crowd-anger of the moment, which both inflamed popular assemblies, are contained by the more professionalized political processes of representative bodies.

However, the loss of direct political input by the citizens of the democratic nation states does engender alienation in these citizens. Distrust of, and cynicism about, politicians abounds in most representative democracies. Of course the citizens can vote out any politician they distrust, but the alienation is not reduced by such an act. For it is the loss of direct participation that creates the alienation, and, it seems that even regular electoral assertions by the citizens are not enough to convince them that their voice is being heard. Even though the citizens' voice is being heard indirectly through electoral processes, the loss of the direct participation still remains problematical.

As the modern nation-states have become mass societies – with the subsequent loss of community and extended family ties – the alienation from direct political participation has become even more problematical. While the limitation of power of leaders and the legal protection of citizens' rights have functioned fairly well in modern democracies, representative participation has become ever more distant from the average citizen.

This development is ironic, since in its inception the participatory sphere was the key defining characteristic of human democratic political activity. It was, after all, the ability of humans to sit down with each other and intersubjectively share their ideas, argue about their differences, and arrive at a collective decision that distinguished humans from other group animals in terms of political process.

Greek democracy, that democracy from which we derive the very name itself, was characterized most centrally by the popular assembly. When one discusses American democracy in its early years – along with the Constitution, Bill of Rights, and the separation of powers – one always mentions the New England town meetings. Yet, we are confronted in the modern world with a structural situation in which citizen's participation has become difficult to achieve and in which alienation from the political process has been occurring. Mass population size, the decline of stable communities, the tendency to rely on technical experts, corporate managers, business leaders, and professional bureaucrats – all of these have combined to reduce the possibilities for democratic participation in modern society.

At the same time, however, the mass media, especially television, have brought the events of the global world directly into the consciousness of the average citizen. The citizens of Japan, Germany, France, Britain and the USA, Denmark, the Netherlands, Singapore, India, and so on, view – seemingly directly – the Berlin Wall come down, the Tienanmen students stand up, the Russian army join the people and bring communism down, Nelson Mandela shake hands with P.W. Botha in South Africa, the poor riot in LA, and so on.

Being able to 'view' the political events of the world gives the modern citizen a sense of immediate involvement with these events. And yet, because the events are viewed through television, they also seem far away, and somehow, not real.

And so the paradox emerges that the modern citizen feels involved and distant at the same time.

But, what of 'civil society' participation – that intermediary sphere of political party participation, interest group participation, and pamphleteering?

CIVIL SOCIETY IN THE MASS TECHNOCRATIC-BUREAUCRATIC CONTEXT

Political Parties

Michels lost all faith in democracy after his study of the German Social Democratic party showed that the party was controlled by a clique at the top. His 'iron law of oligarchy' was for him the proof that mass democracy was really an illusion.[1] Yet, as Guenther Roth has recently pointed out in an essay testing Weber's and Marx's predictions (one hundred years after Marx's death),[2] modern political parties have become less oligarchic, more open to popular participation than ever before.

In Germany itself, the Greens are certainly a party in which popular participation is running high. And, the popularity of the Greens has in turn opened up the Social Democratic Party to more popular involvement. In Britain, though the attempt at creating a Social Democratic Party has failed, the process has engendered a marked increase in participation in both the Tory and Labour parties. Their recent conventions were a model of popular participatory in-put. The Eurocommunists in France, Italy, Spain and Greece have encouraged political participation, which in turn, has encouraged participation in the socialist and capitalist parties.

In Eastern Europe, of course, the euphoria at the collapse of communism has engendered an outpouring of party participation. And, even though this is frightening in terms of the rise of nationalist, fascist and Stalinist parties, it is encouraging in terms of party participation in general.

In the USA 'grass roots' participation through conservative religious movements has almost overwhelmed the Republican Party, while the Democratic Party was engulfed by feminist and civil rights activists and 1960s-style liberals and radicals.

So, party participation has been increasing in the latter half of the twentieth century, and, as we have shown in an earlier chapter, the New Middle Class has been very active in this kind of political participation. And yet, alienation from the democratic process is also increasing – again a paradox. Because even though citizen's participation in party politics has increased, the average citizen does not participate in party politics. And, even though they do not participate in it, they watch it on television every night. Thus, those citizens who do not participate cannot simply withdraw – they are inundated by politics from the mass media. This increases their desire for in-put, not through old-fashioned party channels, but through the media – the channel the average citizen is 'hooked' into. We shall discuss this shortly. First, let us look at the other segment of civil society which modern citizens can participate in.

INTEREST GROUPS, CAUSE GROUPS, AND VETO GROUPS

Pluralist theorists[3] have redefined participation in the modern context to mean participation in and through 'interest groups'. This pluralist description places an intermediate dimension between the citizen and the representative-democratic state – which the Europeans call 'civil society' – and replaces the idea of a democratic citizenry acting as individuals with that of the 'voluntary associations'. Here I do not wish to enter the debate between the Marxists and the liberals as to whether capitalist democracies are controlled by a plethora of countervailing 'veto' groups[4] acting as special interest groups, or by the capitalist rich.

What I wish to highlight in this discussion is that interest group participation has increased in Western Europe and Eastern Europe, has been high in the USA and Canada, and will continue to be an avenue of participation for citizens in modern mass societies.

However, it must be pointed out, that even where interest group participation is high, the average citizen's 'participation' in these interest groups is not direct. The average citizen does not go to the meetings and participate in the political process – even through this indirect route. Rather, the average citizen supports the cause group financially, and only occasionally becomes involved in the leadership roles of the group.

The interest groups themselves tend to become run by professional fulltime salaried executive directors, who like all elite managers, attempt to control policy formulation. Though most interest groups have active boards and large memberships, the average citizen remains a passive participant, with occasional attendance at meetings, and occasional participation in electoral challenges for the leadership of the group. More often, receipt of a newsletter and financial support are the extent of the political participation.

Therefore, again, as with modern political parties, though participation in cause groups seems relatively high, the actual in-put by the average citizen is low. This explains the continuing feeling of alienation of the average citizen in mass democracies. And, again, the intrusion of television on the citizen's psyche increases the feeling of alienation by forcing political events into the consciousness of the viewer without allowing the viewer to actually 'touch', or affect, those events.

I do not wish to minimize the civil society level of participation in modern mass democracies. The cause groups, the interest groups, the veto groups are essential institutions for the proper functioning of representative democracy in the nation-state. Business organizations, labour organizations, environmental groups, civil libertarian groups, religious organizations, gay organizations, and specific cause or veto group organizations are all critical to the proper functioning of democracy, where representative institutions must be influenced.

Having established that participation through political parties and interest groups is still vital, essential, and vibrant in modern society, now let us turn to the new process through which most citizens in our modern high-technology society are actually connected to the political process – television.

MASS-MEDIATED 'DIRECT' DEMOCRACY

It is obvious now, but was unknown fifty years ago, that the effect of television on the democratic political process is profound. The remarkable immediacy that political events gain through media coverage, and the instant national recognition that accrues to candidates who are covered by the media, or are able to buy time on the media, is amazing.

We 'see' politics through TV as if we were really there. We 'know' candidates through TV as if we had met them in our living room. We perceive events through television as if we were participating in them.

When the Berlin wall came down, we felt we were there; when the tanks rolled into Tiananmen Square, we wanted to tell the students to run; we felt as if we were in one of the attack planes during the Gulf War.

Of course, we are not at these events, and we have not met the candidates. The whole experience has been mediated by television: the Gulf War images were carefully censored – most of the killing was not shown, and, the candidates have carefully created a public 'image' which they attempted to project through television.

Therefore, we now have a totally new political process emerging. As citizens, we are able to become more in touch with national and international events than ever before, and these events seem as if they are directly experiential. This is good, in that the modern citizen is certainly a politically aware citizen. Is there anyone, for instance, who did not know that atrocities were occurring in Bosnia? And, this is also good in that most citizens know their leaders, and the leaders of many other countries, through television. Margaret Thatcher, Ronald Reagan, Charles de Gaulle,

Mikhail Gorbachev, Fidel Castro and other world leaders then and since were certainly household figures. In democratic nations, the evaluation of such leaders – positively or negatively – was inevitable, since they were thrust upon us everyday by the television news.

The negative portion of the new television reality we are living in is the distortion engendered by the mass media. We are, of course, not viewing or experiencing the events or candidates directly, but rather the 'image' of the events and candidates projected by the media. What is not shown, and not said is often as important as what is shown and said – as, for instance, the blocking out of soldier killing soldier in the Gulf War. The mass media, sometimes consciously, sometimes inadvertently, put their own slant or spin on news events, slur or pay homage to certain leaders and candidates, denigrate individuals, nations or political ideas, while paying homage to others.

Sometimes television dominates an event, making its coverage, its anchor people, its media world, more important than the event itself – so it is that in this strange new mass media world emerging, that media 'stars', from television, movies and pop music often become more popular than political figures, and sometimes political figures in their own right! The Ronald Reagan presidency should not be seen as an aberration in this regard, but rather as one of the typical trends emerging in our new mass-mediated world culture.

Thus, the political processes emerging with the advent of television have created the potential for a new form of democracy. However, television democracy is mediated, and therefore subject to media distortion and media manipulation.

Further, television democracy is beginning to exhibit three basic features: (1) the experiencing of political events, political leaders, and candidates for electoral office through television exposure, almost exclusively, (2) the desire of the citizens to 'interact' through the media themselves – especially television and radio – with leaders and candidates, and 'participate' in issues debates. And (3) the desire by modern citizens to by-pass their elected legislatures – both local and national – through direct referenda on key issues of the day.

This combination of television connectedness to events, and leaders, candidates and issues, interactive talk-back by viewers, and instant referenda on local and national issues, may produce a new form of democracy, very different in structure and process from parliamentary democracy. We are calling this new form of democracy, mass-mediated direct democracy.

Let us look more closely at the interactive potential of television, and the instant referenda potential of television–computer plebiscites.

ACCESS TO THE MASS MEDIA

These days a candidate isn't 'real' unless he or she is seen on television. But, to get on television, the candidate must raise – or inherit – vast amounts of money. Television costs have ballooned as 'image makers' and 'slogan slingers' intervene to 'package' candidates and 'market' their ideas. If you are not on television, you are not a visible candidate. Pressing the flesh at subway stops and shopping centres is still necessary, but somehow if the candidate is not a media star, the excitement of the handshake isn't there.

Campaigns are more and more determined by the mass media – especially television. Your television image, your television message, and the television interpretation of your candidacy is what reaches the majority of the voters.

If you can't afford television, your chances of winning are greatly diminished. Just by purchasing television time a candidate becomes viable. Anyone rich enough to buy television time can buy legitimacy. Such candidates don't always win, but the percentage of votes they have garnered – given their unknown political status prior to their television campaigns – has been remarkably high. Conversely, candidates without media time are having increasing difficulty getting their message across to the voting public. In fact, their very credibility as candidates comes into question.

The system should provide for another way to finance US elections, and it should also provide for fair access to the mass media. There is a better way: a menu of reforms that can work – because they have worked, elsewhere.

First, Congress should enact the 'Clean Campaign Act'.[5] It would set up matching public funds for candidates. For instance, if a candidate limited him/herself to $70 000 in special interest gifts, the government would provide up to $90 000 in matchable money. The money would come from a voluntary taxpayer checkoff. This has worked in states like Wisconsin and for the presidential elections.

With such a system, vested interests could still make their legitimate influence felt, but elitism and plutocracy would be inhibited. The democratic process must remain open to all citizens – as voters and as candidates. As we approach the year 2000 let us be sure to institute reforms that extend, rather than restrict, our democratic participation.

Second, the US should guarantee *bona fide* general election candidates a minimal amount of radio or TV time, to ensure that their message can be delivered without special interest cash. Too expensive? Citizen groups have estimated the cost to come to just a dime per citizen. It has worked in Britain and France, so why not for the US?

In France, major candidates are guaranteed free television time,[6] while in the State of New Jersey, candidates receive state funding for use on the media. Whether free or government-financed, equal access to the mass media has become as important for democracy as 'equal protection before the law'. Modern democracies should propose and enact a fair media access law in the very near future.

Media access for candidates is crucial, since most modern elections are carried out through television. However, the potential of television goes far beyond the campaign process.

INTERACTIVE TELEVISION TOWN MEETINGS

Right now the interactive potential of television (and radio) is limited to the 'talk show' format. Viewers (listeners) can call in by telephone and make their opinions known or argue with the hosts or guests on these shows.

However, the future potential for interactive television will be greater. New media systems are already being experimented with that combine television with computers and telephone lines. Eventually a fully interactional media system may evolve.

For instance, transnational corporations are already able to engage in teleconferencing. These corporations have interactive television systems through which individuals in Hong Kong can talk with individuals in London or New York, or Toronto, etc. These teleconferences are fully interactional in that both voice and video image are clear and immediate. One sees this also on American news shows, where the Washington correspondent speaks with the correspondent in Jerusalem or Zagreb as if they were chatting with each other in their home studio.

How could this become a new political institution?

If every household eventually contained an interactive television system, then local and national town meetings could be held in the same way that teleconferences are now held. Of course, because of the larger size of electoral districts, only some citizens could actually be seen and heard during a television town meeting. The choosing of citizens to be seen and heard could be accomplished on a first-come, first-served basis, as with TV talk shows, or small electronic districts could be created, with each district taking its turn on successive days and time slots.

Because of the mass population size of modern societies, even local districts tend to have large populations, and therefore the electronic districts might have to be small for real teleconference style interaction to occur. However, with the addition of the computer to the interactive television systems, every citizen could have their opinion registered through an instant-computer tabulation vote. We shall discuss this shortly. Further, television telephones are already being perfected, and these could be hooked through cable to political channels on television.

As John Wicklein describes it, all modes of communications we humans have devised since the beginning of our humanity are coming together into a single electronic system, driven by computers. The focus of the system will be a home communications set (HCS) that looks like a standard television with a keyboard attached. Within it will be a small microprocessor that turns into a computer terminal as well. The information and entertainment it provides will arrive over optical fibre 'wires' of glass or from a communications satellite sending signals directly to small disk antennas on our rooftops. Later models will include a small television camera and a microphone, to make possible video telephone calls. The set will give us news in print as well as the usual video form; it will display either text or pictures on its television screen. If we want hard copies, it will produce them quickly on an attached printer or photocopy device.

The set will be able to supply hundreds of channels on television. These will be used for standard commercial programming, service programming for special interest groups, educational programmes, or electronic catalogues on 'shopping' channels.

The most important feature of the new communications system will be its capacity to be two-way, permitting us to respond over the system to what the system is offering us.

Already in existence, there are two variants of the interactional television systems. The first variant contains the television and the computer system, but does not include the video camera and microphone which would allow for complete interaction from the home to the head end, or media center, of the cable system from which the programming emanates. It does allow partial interaction. That is, through computer buttons, the person at home can register his or her views in a simple yes, no, undecided, or multiple-choice fashion.[7]

The second variant allows the individual in the home to interact with computer buttons, or with full capacity video and audio two-way telemeetings. That is, the persons at the head end could, by selecting your home (in turn) see you and hear you, as well as you could see and hear them.[8]

This would operate like a phone call to a radio station operates now. The only impediment would be that if hundreds of home two-way equipped citizens wished to interact with those at the head, or centre, only some could do so. Who would be chosen and in what order would be problematic.

However, through the computer system, everyone's point of view would be instantly tabulated and included.

At present, most interactive telecommunications sets and systems have only the computer interactive capacity. However, television–telephone–computer–cable systems are just now being developed. And, already there are some experimental systems with more complete interactive teleconferencing capacities in existence. In Japan, the USA, Canada, and other high-tech nations, such systems are being experimented with in terms of their democratic potential. The 'Qube' system of Columbus Ohio and the 'High Ovis' systems of Japan are typical in this regard.[9]

The Qube system at Columbus Ohio is an example of the first variant. It has no home video camera system. Therefore, those at the head end, or programming centre, cannot see or hear those at home. However, those at the head end can receive computer messages on all questions presented, and can determine the percentage of the citizenry whose sets are tuned in.

Participation is developed by the system in the community channels, where viewers are solicited to 'interact' with their sets. Most of the inter-action centers are on the 'Columbus Alive' channel.

A 'polling' computer, it gathers billing and response data from sub-scribers. The computer sweeps all subscribers' homes at six second intervals, asking: Is the set turned on? What channel has been punched up? What was the last Response Button touched?

The home consol has five Response Buttons in addition to the channel selection buttons. The first two can be used as 'Yes-No' buttons; all five can be used to answer multiple-choice questions or punch up number codes to indicate, as one possibility, selection of products displayed on the screen.[10]

The Qube system regularly runs social surveys and political referenda. For instance, one evening a social survey was conducted on homosexuality. 'If you know a homosexual,' said the host, 'push button #1 for yes; if you do not, press button #2 for no. Touch in now.' Within seconds, the computer supplied the results: yes – 65 per cent; no – 35 per cent.[11]

The Qube system is also used for commercial transactions. The impact of this system could be monumental. Banks and shopping centres may become obsolete – although in the case of the latter, people may prefer to go shopping – as a recreational activity.[12]

This is real interactive television – town meetings can be held through the Qube system. No video cameras or microphone are included in the home set as yet. However, by responding through the computer key pad, home viewers can definitely make their views known to the political leaders (or commissioners) at the head end. The computer–television town meetings have been very successful in Columbus, and, they have stimulated citizens' participation in the political process in general.[13]

The Japanese have introduced two experimental interactive telecommunications networks. One is called TAMA, which is more like Qube, the other, Hi Ovis, which is the most highly developed two-way system in the world. Hi Ovis may be the wave of the future.[14]

Hi Ovis tested what was considered the 'next step' in interactive television systems: an audio and visual return capability provided by a camera and microphone built into the HCS (home communications set). This made it possible for the people at home to be seen and heard throughout the system when they participated in a discussion of community affairs. They could also indicate their reactions via computer key pads.[15]

The Hi Ovis system has been remarkably successful. Through the computer, as in the Qube system, everyone's views can be instantly surveyed and tabulated, while, through the two-way television and audio capacity, individuals can 'phone in', or see and be heard by the politicians or government officials or anybody else at the studio end of the system.

Because the Hi Ovis constituency is small, everyone who wants to phone-in and be seen and heard usually can get a turn to do so. Again, where constituencies are larger, this would become difficult.

The question of whether the constituencies themselves, or local politicians, or commercial cable operators will control the content and programming of the citizen's access channel becomes central here. Obviously, the citizens, in consultation with their elected officials, should determine the content of the town meetings. And, when it comes to town meetings, the commercial cable operator should only be involved with the technical end of the process. Since such commercial cable operators will be making huge profits from the regular programme channels, they can be convinced, and should be convinced, to provide the technical expertise for the establishment and continuing operation of the public access democratic-participatory technological town-meetings.

What a crossroads we face! Will the new telecommunications technology lead to Big Brother, Technological Democracy, advanced electronic commercial economics?

Will the new communications collect information from the masses and funnel it to the few at the top, tending to centralize government and industry, or will it be a decentralizing force, providing more information to the populace so that more and better decision-making can be done at the local levels?[16]

The situation in each nation is different. Therefore, citizens in Britain,

France, Germany, Scandinavia, Japan, and so on will have to appraise the situation and act to establish public access channels in their nation. It would seem to me that given their traditions, Britain and Sweden might pioneer in this process – while Germany often follows Sweden in such matters. The situation in France, where the government has often used television in a dictatorial way, may be problematical, while the situation in America, that most democratic and most capitalistic of modern nations, is such that it may yield its democratic tradition to its commercial one, if its citizens do not act quickly and forcefully.

At this moment in history it is clear that computerized information systems will occur, including electronic mail and library services; television–credit-card purchasing and bill-paying will occur; corporate teleconferencing will increase; commercial entertainment channels will proliferate; and, unfortunately, computer dossiers and telecommunications surveillance will expand, increasing the potential for technological totalitarianism to occur.

What is not clear is whether public access channels will become institutionalized, so that TV town meetings, referenda, community affairs programming, and C-Span (Parliament viewing) programming will enhance the potential of a modern form of democracy.

Let us look further at this democratic potential.

ELECTRONIC POLITICAL DISTRICTS

One of the problems for democracy in mass society is that communities, as we once knew them, have declined. People, often neighbours, do not know each other. New suburbs grow, while older neighbourhoods decline. This is a world-wide phenomenon, although because of the continental size of the United States, its frontier tradition, and massive in-migrations, the situation is more dramatic.

Democratic politics is based on community districts. Town meetings and congressional districts assumed relatively stable community conditions. If the modern individual is barely part of a community, that individual will hardly be likely to attend community meetings or be aware of elected representatives.

Democratic communities have changed over the course of time to fit the new structural circumstances of changed societies. Among primitive bands, the entire band was the democratic community. However, among societies as small as tribes, already the democratic community was subdivided, and the clan was the key local unit, while the tribal council was the 'national unit'.[17] Greek city-states had electoral (and lot) districts called 'tribes', though these were not tribes, but electoral communities (the Greek word for tribe was *ethne*; they used the word for electoral grouping). And, of course, parliamentary democracies have electoral districts based on geographic and demographic characteristics.

Given the communityless condition of modern mass society, new districting approaches linked to the telecommunications media have become necessary. The geographic and population size districts can be maintained, but they should be supplemented with the narrowest cable channel capacity of the television–computer–telephone–home-communications-set system.

Citizens would not have to know their neighbours in order to participate politically through the television–computer hookup. Furthermore, new residents – and there will continue to be a great deal of moving – would be able to gain a knowledge of local public affairs issues, local representatives (who would appear regularly on the public affairs channels), and the national issues of particular concern within the district. The newly-arrived citizen could almost immediately become part of the political community created by the telecommunications media.

Given the homogenization of national and world cultures which is occurring (in part because of the mass media) the new resident will undoubtedly find it increasingly easy to feel part of, and participate in, new communities. In fact, if interactive cable programming – including public access channels – becomes nationally and internationally institutionalized, then it will become easy to become an active citizen in a new political community. Like blue jeans and rock music, television town meetings may become world phenomena, taken for granted as a part of modern life – except where they are repressed. Already, for instance, the Internet has citizens from every nation in the world talking to each other. Such citizens not only share the technology, but are beginning to share certain cultural orientations to the world.

ELECTRONIC INTEREST GROUP COMMUNITIES

If actual communities have declined in modern society, interest groups have increased. These interest groups are not actually groups, but they are national or international organizations to which people 'belong' through supplying money or through attending occasional meetings, programmes or rallies.

244 *The New Middle Class and Democracy*

Interactive cable television could revolutionize interest group politics. That is, a number of cable channels devoted to interest groups and with interactive capacities could be instituted on a national, or even international, basis. Therefore, no matter where a modern citizen might move, he or she could have instant contact with the interest groups which reflect his or her political proclivities. The ability to participate with the interest groups of one's choice would be enhanced in two ways. First, anywhere in the nation (or perhaps world) that a person moved to, they could maintain contact with their interest groups. And, second, the citizens could interact with the managerial and elected officers of the interest group, expressing their views through computer referenda or two-way discussions.

In this way, the citizen's contact with these important intermediate political groups (intermediate between the citizen and legislature) would become more immediate, and the citizen's in-put more direct. The massification of society and decline of community would be minimized in its political effect, and civil society enhanced by the new media.

Political constituencies, in the form of interest group town meetings would supplement the geographical community town meetings. Today's citizen is often more connected to interest groups, cause groups, and veto groups then to his or her local community politics. This is good democratic politics, and therefore, the new telecommunications media will be as important for the maintenance of democracy in modern society, as will the enhancing of the geographical communities' democratic potential.

ELECTRONIC CONTACT WITH ELECTED OFFICIALS

In today's mass society, many citizens do not know who is their representative – national or local. Through interactive television this could be remedied. The elected officials of Reading, PA and Columbus, Ohio, regularly appear – at the head end, or programming centre–on the interactive cable channel. Once systems like Hi Ovis in Japan become institutionalized, question and answer sessions between citizens and their elected officials may become commonplace. The citizen would not have to leave his or her living room and could still make meaningful contact with elected representatives. Through interactive television, the elected officials might really become the people's representatives in a very deep sense. Of course, citizens can vote-out any elected official they disapprove of. However, through cable television, they could really get to know just what policies and programmes the representative was pushing or opposing.

COMPUTER–TELEVISION–HOME VOTING SYSTEMS

Voting, in the near future, may be accomplished right in one's home. Television–computer–cable hookups may allow us to vote by way of a computer key pad linked with a cable television channel.

This could increase voter participation, but it could also be open to fraud or electronic tampering. Let us assume, however, that the electronic home voting system was perfected, and that it was adequately protected against electronic manipulation.

If this system were in place, then a whole new form of direct democracy could emerge from it. For, not only could candidates for office be viewed and voted upon through the mass media, but issues could be aired and voted upon as well.

Computerized instant referenda on key political issues could occur, catapulting us into a new age of mass-mediated direct democracy. Let us look closely at this.

COMPUTER REFERENDA: DIRECT DEMOCRACY AT THE LOCAL AND NATIONAL LEVEL

California often sets the trend for the USA, and the USA often sets the trend for the world. In California, referenda on hot political issues have become, perhaps, more important than the state legislature. Furthermore, the California referenda have exhibited all the flaws that conservative-democratic theorists warned against in terms of direct democracy, and all the excitement and involvement that more radical-democratic theorists promote.

What have the California citizens been voting on? Taxes, immigration, race-quotas, gays and lesbians – inflammatory issues, which conservative-democratic theorists believe are best left to the cooler heads of the elected representatives than to the 'mob-rule' of the citizens themselves.

Not only in California, but also in many of the Western states, issues referenda are becoming popular. When such referenda touch 'hot' issues, they really become popular. When they touch dull issues, like sewage system construction, few citizens vote on them. But when the issue is something like abortion or racial hiring quotas, huge 'hot' voter turnouts occur.

The referendum system can be dramatically increased, if such referenda are conducted through home computer voting outlets. Then we would have an 'instant referendum' system – the ultimate in direct democracy. Notice, such referenda by-pass the legislature and truly create a new form of direct democracy on a mass-media base. Let us look at the positives and negatives involved.

FOR LOCAL AND NATIONAL REFERENDA

The possibility for local and national referenda to occur as a routine political process will emerge as the new computer–television–telephone technology becomes perfected and becomes available to the majority of the modern citizenry.

However, the political participation value of such referenda must be carefully weighed against the dangerous potentialities lurking within it. We have already suggested that in the California referenda, hysteria, mob psychology, fear and bigotry have already displayed themselves. That is, citizens have voted to limit taxation, engendering cut-backs and deteriorations in services that they later regretted. Citizens have voted to exclude immigrants – targeting certain racial and ethnic groups by specific discrimination. Californians have also voted to remove the minority-preference quotas that were helping blacks get ahead. In Colorado referenda against gays have been held and passed.

All of these issues are volatile. And, similar, hot issues could be raised in European referenda, especially focusing on immigrants and taxation policies.

Direct democracy through referenda can be a volatile process, then, inflaming conflict rather than reducing it. With the mediating effect of the legislature removed, the crowd psychology and demagoguery of the moment can break through, making of democracy 'mob rule',[18] instead of legal-rational rule.

Still, there are those theorists, such as Benjamin Barber,[19] who make a strong case for direct democracy. In his book, *Strong Democracy*,[20] Barber presents a spirited argument in favour of increasing amounts of direct democracy, as linked to the new computer–interactive television systems.

Barber, an expert on the Swiss referendum systems – which is a national system – describes precisely the way in which such a national system could be utilized in the larger democratic nations. Barber is familiar with the critique of direct democracy, and describes the ways in which safeguards are built into the Swiss system.

These structural safeguards have been developed in order to prevent the Swiss from arriving at hasty decisions, or giving in to the hysteria of the moment. They include long periods of reflection in which repeat referenda must occur two or three times before the referendum vote is considered really ratified and binding.[21]

Barber argues for a national referendum system which would include a mandatory tie-in with neighbourhood assemblies and inter-active television town meetings. He also suggests – as in the Swiss system – a two-stage format, so that the people get a chance to change their mind, or to affirm their first vote. In his programme, there would be a waiting period of six months followed by a second vote. Congress (Parliament) in his system would have the right to veto the vote, but the people could vote a third time to override the congressional veto.[22]

Since the intention of the process is to increase participation, rather than to produce immediate legislation, the deliberate, even ponderous pace, of a two- or three-step process would be justified.

In terms of public information on the issue in question, Barber suggests that informational documents offering pro and con arguments on each issue would be made available to the people.[23] Such information already is provided by such organizations as the League of Women's Voters, in a non-partisan fashion, on current ballot initiatives. Cable channels could also be utilized for these non-partisan pro and con summaries. Interest groups, of course, will fund partisan messages. They often inundate the media with their position. Those with more money gain far greater media time. Therefore, Barber would have the law limit the spending by interest groups. This latter, however, has never been successfully accomplished in the USA; one-sided media 'blitzes' have often occurred with fairly good success. However, television town meetings, if institutionalized, could counteract interest group media blitzes.

Other theorists suggest that computer-television referenda on issues should be conducted, but, they should be non-binding, more like a poll than a vote. The idea is that such non-binding referenda would serve as a guide to the legislature, thus increasing the level of direct democracy without by-passing the legislature. The people, in turn, could vote the legislature out, if it did not follow their will. Therefore, even in a non-binding form, a system of instant-referenda (through computer–television systems) could add a dimension of direct democracy missing since the days of the Athenian's popular assembly.

AGAINST LOCAL AND NATIONAL REFERENDA: CONTRA MOB RULE AND CROWD PSYCHOLOGY

Theorists from Polybius[24] to Madison[25] have warned that representative democracy may be more stable than direct democracy. The people – in an instant referendum situation – could be swayed like a 'mob' through a campaign of propaganda and inflammatory rhetoric, for or against an issue. This is why Barber suggests the Swiss system, which requires a number of separate votes over time.[26] But, even given the Swiss system, a process of binding referenda which by-passes the congress or parliament could be problematic. It would create a direct democracy, and this is exciting, but it would also increase democratic excesses and instability – minority rights could be trampled upon, hysterical solutions could be foisted upon the populace by demagogues. These kinds of excesses have occurred in California already. The debate between the Madisonians and the Rousseauians has already been revived.

In my view, the worry about an increase in direct democracy is wonderful. For, just a few years ago, we could envision nothing but despotic utopias, like *Brave New World*[27] and *1984*[28] – the 'iron cage' and 'Big Brother' were all that were waiting for us. Now, with the new telecommunications media, we are arguing over 'instant referenda' and an increase in 'direct democracy'.

Though we ought to worry about mob rule, we should not shrink from extending the process of democracy – as long as this process is carefully constrained by constitutional legal parameters, and the referenda include some of the Swiss safeguards.

COMPUTER-TELEVISION VOTING AND POLLING

The new communications systems will become commonplace in the modern home. When the systems are in the majority of homes, voting and polling through them will become as common as banking at a machine. Mail service, library service, and Internet international communication processes are already expanding. Therefore we should predict that voting machines and ballot boxes will be replaced by home computer-television voting.

Stump political speeches have already become less important than television campaigning. And, soon, the candidate vote and issues votes will be conducted through television-computer systems. Campaign polls will become more accurate as well, instant polling data becoming available through television-computer polling programmes.

Since these new telecommunications systems will come into the political process, should we not carefully guide them into place? We must make sure that the safeguards of the Whig-liberal traditions are not denigrated or discarded. The fears about direct democracy are real, as is the greatness of it.

Pericles' funeral oration still brings tears to my eyes and urges me toward citizen's participation in democratic politics. However, Alcibiades' cynical manipulation of the Athenians and their subsequent ruin at the battle of Syracuse also bring tears to my eyes. Direct democracy is too volatile, can run roughshod over minorities, has produced irresponsible mob-like decisions. But, representative democracy, in a mass nation-state, becomes too distant from the citizen, engenders alienation, and discourages citizen's participation.

With the new telecommunications technology, we may be able to create a better balance between representative and direct democracy. If we do it right, we may re-create the popular assembly through interactive television and computer systems, while retaining the constraints of constitutional law and the deliberate cool-headedness of the representative processes.

DIRECT DEMOCRACY AS A COUNTER-FORCE AGAINST BUREAUCRATIC DESPOTISM

As a counter-force to bureaucracy, media direct democracy may be a necessary ingredient for modern democracy. If decision-making has come to reside more and more in the power elite of corporations, government bureaus, and the military – as Mills insisted – then perhaps the increase in local and national interactive television participation can counteract this kind of elite decision-making, and the alienation it produces. Through electronic direct democracy, citizens can act on issues that affect them, without waiting for the power elite or the legislature. This makes democracy real again, and puts power back into the hands of the citizens and out of the hands of distant elite managers and bureaucrats

TECHNOLOGICAL TOTALITARIANISM

'1984' need not occur. Yet, the technology available to foster totalitarianism is awesome. Already surveillance devices, computer dossiers, sophisticated policing, weaponry and torture devices make police-state despotism and 1984 totalitarianism clear possibilities.

The increasing sophistication of such devices should encourage us to increase our control over them in parallel fashion. Civil libertarians in the United States have already been warning that government and corporate surveillance should be limited and monitored by the Congress and the courts; that FBI, CIA, Pentagon and IRS (tax bureau) activities in this regard should be carefully controlled, perhaps even restricted; that corporate spying and corporate intrusions on the personal privacy of their employees should be stopped entirely.[29]

In Scandinavia and Germany, the ombudsman has been empowered specifically to monitor military, police and spying activities while, in England, a growing concern over these processes is stimulating various proposals for their control (from legal restrictions, to ombudsman institutionalization).

Democratic nations with strong legal traditions or with ombudsman traditions have debated the growing use of communications technology against their citizenry and have even begun to limit such usage. In despotic nations the expansion of the use of advanced technology in an authoritarian direction had rapidly occurred, but not without protest from at least portions of the new middle class within these nations.

In both democratic and despotic nations, however, due to the principle of competitive advantage, and also for reasons inherent in the structure of mass society, the adoption of communications technology for despotic purposes has occurred. In terms of 'competitive advantage', what I am referring to is the use of spy systems, like the CIA and KGB, and the use of corporate spy systems – if one nation has one, the other must have one, and the same goes for competing corporations.

In terms of the structure of mass society, however, there is no doubt that in modern, communityless society, terrorist groups, organized criminal groups, and revolutionary groups could render the society helpless in the face of their use of modern technological violence unless government surveillance of such groups is sophisticated enough to inhibit their violent activities.

SURVEILLANCE IN DEMOCRATIC NATIONS

There has been a dramatic increase in the amount of surveillance against citizens emanating from government and corporate organizations in modern democratic nations. Why can't we just do away with it – make such activity completely illegal? The reason is that a democratic mass society would then be helpless in the face of certain domestic and international threats. These threats would emanate from organized criminal groups, revolutionary groups, terrorist groups, socio-pathically 'insane' groups, and 'enemy' nations. The militia movement in the USA is typical in this regard.

Because of the ability of such groups to obtain modern weaponry and communications technology, such a group could become threatening, not only to individual citizens, but to an entire society. Therefore, no modern society can exist without the necessary institutionalized power to control such groups.

The maintenance of order has always been the purpose of government. Hobbes[30] first formulated this postulate, and Weber[31] further analysed it by describing 'the state' as holding a monopoly on the legitimate means of violence. However, Locke[32] debated Hobbes on just how much power needed to be ceded to the government by the citizens in order to maintain order, and we, as modern citizens, should debate Weber on how much violence should be ceded to the modern state.

How much power is enough power to control modern anti-social groups (or foreign nations)?

If we study the American FBI we will see that this national government police bureau became necessary when criminal activities and criminal organization grew in power beyond the ability of local police agencies to control them. Further, the scope of the FBI was expanded to include control of political groups of the left and the right, black movement groups, then terrorist groups, and deviant or socio-pathic groups. Some of these groups were violent and armed, some were not. Some were seen as socially disruptive by the majority of citizens, some defined as disruptive by specific elements of the society, and some by the FBI itself.

The FBI does not eliminate organized crime, terrorism, revolutionism, or deviance, just as local police forces do not eliminate local crime. National and local police forces do, however, allow government – with the consent of the citizenry in democratic polities – to contain and control national and local anti-social violence and to react actively against such violence when it occurs. Few citizens of democratic societies would want to eliminate the police, though civil libertarian protection against the police is absolutely essential in order for democracy to survive.

THE DESPOTIC POTENTIAL OF THE NEW TELE-COMMUNICATIONS SYSTEM

An interactive system which supplies us with most of our information and entertaining programming, delivers our phone messages and our mail, carries out all our financial transactions, and even senses movement in our homes can be used to invade our privacy and order our activities. The biggest threat of a multifaceted, integrated communications system is that a single authority will win control of the whole system and its contents.[33]

Computerized dossiers are already being compiled by government and private organizations with very little public outcry, because most modern publics are unaware of the danger at this present time.

Computerized 'services' make it possible to develop detailed personal files on every one of us that could be exchanged electronically between organizations collecting data. With a computerized data bank tied into the home communications set, such dossiers could be made available to anyone who wished to pay the service charge. Electronic surveillance via computer of phone conversations and written communications could then add to the completeness of the dossiers being assembled.[34]

The surveillance capacities of the new telecommunications systems are already developed to 'Big Brother' capabilities. 'Spook' devices now exist for switching on a private telephone from a remote location and amplifying the sound it captures sufficiently for pickup of conversations across a room. The same techniques could switch on the camera and a microphone in the HCS, making possible video and audio surveillance of the home without revealing this surveillance to the residents. The capability now existing makes George Orwell seem to have been extremely prescient when he described in 1949 the telescreen's constant monitoring of Winston's presence in his own home in the mythical Britain of 35 years later.[35]

The frightening part of the surveillance capacity of the new technology is that it may be allowed to be introduced through two seemingly legitimate sources. First, through the extension of such 'routine' procedures by government and corporate agencies which already use bugging devices and other surveillance techniques as part of their everyday operations. The FBI, CIA and the corporations may simply extend their capacities as the technology improves. This process may emerge so gradually that no limit is set upon it.

Intrusions gain legitimacy, even in a democracy, through the placid acceptance of rules, practices and customs promulgated not only by government, but also by commercial 'service' companies.[36]

Second, surveillance may become institutionalized inadvertently through the installation of burglar sensory devices, television medical-diagnosis systems, and two-way television and telephone systems. These latter would be installed for safety reasons, humanitarian reasons, or for increased communications potential, but could become perverted into the totalitarian device *par excellence*. The way is being broken for in-house surveillance by seemingly benign 'advances' in two-way service. In France, Britain, Australia and parts of the United States, two-way systems are being used to read electric, gas, and water meters. But once such sensing devices are installed (the two-way device that senses the motion of a 'burglar' in your house is a case in point), the background system has been laid for unauthorized monitoring by government or private investigators.[37]

How can we protect ourselves from the despotic potential inherent in the new communications technology?

Sweden has pioneered in the attempt to protect the citizen from technological totalitarianism in the same way in which it has pioneered in the protection from bureaucratic authoritarianism (with its ombudsman system).

In the United States, the FBI is already routinely using telecommunications devices for surveillance. Further, private companies, such as Equifax in Atlanta, Georgia, collect data for insurance and other corporations. Aside from FBI dossiers, which would in most cases be expected to be damaging to a citizen, the dossiers of the private companies (or the tax bureau or any other government or corporate organization) could be intentionally or inadvertently damaging.

The Swedish Data Inspection Board is another ombudsman device. It advocates for the people against government and private computerized data systems. For instance, when Stig Bjererstrom was falsely accused by the government taxation department's computer of failing to pay his taxes properly, he appealed to the Data Inspection Board. This agency, which regulates every computer data bank in the country that collects personal records got right on his case.[38]

> The Data Inspection Board has the power to . . . direct the keeper of a computerized file to correct or drop false information it contains to an individual. The power covers not only government agencies such as the Taxation Board but also private concerns such as consumer and credit investigation services. . . . Even Intelligence Service files are no exception. . . . In 1972 Sweden became the first country to pass national laws against the threat from computers in both the public and private sectors. . . .

In 1972 it enacted laws to permit citizens to get at their records even when these were in the form of electronic bits stored in a computer. It gave them the right to see printouts of files kept in any data bank and make corrections in these files if they were in error.

> Every citizen has the right, once a year, to inspect any computer file that is kept on her or him and to have a written printout of that information . . . [Any] corrections must be sent to anyone to whom the information has been transmitted previously. If the record keeper refuses to correct the file, the citizens may appeal to the Data Inspection Board.[39]

Other nations have adopted or are debating the adoption of the Swedish system.

The Swedish experience with data protection had a strong influence on other Western European countries. By 1980 Denmark, Norway, Finland, and Austria had set up data boards. In West Germany, the states established data protection agencies and the federal government appointed a 'data ombudsman'.

> France placed privacy protection under an independent agency, the Commission Nationale de L'Informatique et des Libertés. . . . In most of its provisions, the French act follows the Swedish precedent. . . . [However] if a dataholder refuses to comply . . . the citizen of the Commission must appeal to the courts for enforcement.[39a]

The French situation may provide the precedent for the United States and Britain. For in the countries with a long-standing legal tradition, the courts, rather than the ombudsman institution, may be the more acceptable road to control of computer dossiers. Either road to the control of the despotic potential within the new communications technology is acceptable. Both cultural traditions could be made effective in this regard. Within countries such as Sweden, with a tradition of ombudsman agencies, a clear set of legal rights protecting the citizens from the new technology should be established. Within the United States constitution, and within the European Charter of Human Rights, a new Bill of Rights relating to the citizen's right to privacy from, and protection from, the 'Big Brother' potential of telecommunications should be carefully spelled out and enacted.

12 Education for Democracy in the High-Technology Global Village

UNIVERSAL EDUCATION

A nation that wishes to establish and maintain democracy must create an educated citizenry. This means that such a nation attempts, as far as it is able, to educate all of its citizens. Universal literacy, through universal grade-school education, becomes the first goal. Once this is established, then the level and depth of schooling must be extended in an ever-increasing process.

In the contemporary world, a universal college education will become the norm. As difficult as this is to achieve, the world of the near future may demand even more: a universal graduate school education. A master's degree is already a prerequisite for most modern occupational roles.

What of educating the poor? In most nations, this is a difficult problem. Whether the poor are urban or rural, if they come from a non-literate tradition or from illiterate parents, their education becomes a more difficult task to accomplish. We know, today, however, that if children are brought into the educational process at a very young age – from 2–3 years onward – they can do better than if they begin at 6 or 7 years of age. Given this knowledge, a world programme of pre-school education could go far in helping to raise the literacy level, so that the universal education system will better succeed. Such pre-school programmes can be as effective in the villages of the Philippines or the Andes as in the poverty neighbourhoods of New York or Liverpool.

Of course, we also know that among peoples where education is greatly venerated – such as among the Chinese, Indians, Greeks and Jews – success in universal education will be greater. And, where family breakdown has occurred – as among the poor Brazilians – education will be more difficult to achieve due to lack of nurturance, inadequate socialization, and limited vocabulary. Nonetheless, pre-school educational programmes could go far toward equalizing the educational potential of all children.

What of elite schools for the rich? In a democratic society, equality of opportunity must be established. Since a strong education is one of the bases of equal opportunity, discrimination based on wealth is problematical. A society attempting to stabilize democracy strives, as far as it is able, to provide equal educational opportunities for all of its citizens.

Elite domination by the rich – fashioned through control of an exclusive educational system – could create resentment and hostility among worthy citizens of lesser means. This was the case, for example, in Peru and Mexico, and also in Britain until after the Second World War. As Aristotle cautioned, a careful balance between the privileges of wealth and the opportunities for the general citizenry must be maintained in order to stabilize democracy. It is hoped that, in the long run, in any democracy there will be equal opportunity to gain the best education possible. In this way, both the wealthy and the general population will be able to contribute – to the pinnacle of their creative abilities – as citizens of the nation.

Today's 'knowledge elite', in any case, is no longer coextensive with wealth. In the modern world, knowledge is linked with high-tech specialization and academic talent. This is all the more reason for modern societies to move away from elitist educational institutions linked to wealth, and move toward open educational systems based on academic merit.

EDUCATION TO THE RATIONAL-SCIENTIFIC WORLD-VIEW AND TO LEGAL-RATIONAL AUTHORITY

It is not just an education, or even a university education, which undergirds and safeguards democracies. We have seen instances in history where well-educated individuals or groups preferred elitist forms of government – perhaps Ortega y Gasset[1] most forcefully argued for Platonic elitism in the modern world. Therefore it must be made clear that it is not education in and of itself that creates the 'democratic ethos'. After all, since the days of Sumer, Ancient Egypt and Ancient China, scribes and other officials have been educated meticulously in mathematics, linguistics and literature – and have made great strides in the accumulation of knowledge in many fields.

The officials of the ancient world were also, however, educated to a magical-religious world-view, which projected nature as controlled by gods and spirits and demons, arranged in power hierarchies. From this, political conceptions flowed, investing divinity in the kingship, magic in the priests, and a hierarchy of authority among humans.[2]

All of this is well known, but it needs to be emphasized, because within such a 'traditional' magical–monarchical world-view, the best education a nation could provide could not create a democratically oriented citizenry. Therefore, the content of universal education becomes important for the sustenance of democracy.

The rational-scientific world-view makes central the basic belief in free inquiry, the 'lawfulness' of the natural world, and by analogy, the lawfulness of the social world. The rational-scientific world view also wipes away notions of magic and spirits as causal forces, and engenders a 'disenchanted', but rational individual.[3] With magic gone and the spirit world benign, priests and kings lose their 'aura', and the hierarchy of officials attached to them become – instead of venerated mandarins – annoying 'bureaucrats'. The teaching of the rational-scientific world view, then, becomes part of the foundation for democracy.

This is not to say that the rise of rationalism and science makes religion irrelevant. For, if the divine is stripped of its immanent causality in our everyday lives, it is still sought for as a 'first cause' and mystical presence, in a seemingly inexplicable, infinite, universe.[4] And if morality loses its sacred quality, as Durkheim described it, the study of ethics becomes all the more prescient, as modern societies flounder in a malaise of anomic confusion, and strive to establish a humanistic ethics on a more rational base.[5]

However, if science wipes away magic and ritual, it reinforces the idea of 'law'. For not only is nature viewed in terms of laws – rationally investigated – but so too is society conceived in terms of a system of rational laws which guide human actions.[6] Rational-legal authority,[7] to use Weber's terms, becomes the proper regulatory mechanism for a democratic society. Constitutional law acts as a guide and constraint upon government power, and, judicial law becomes the proper forum for the rational determination of criminal justice.

Now, having established that the instilling of the rational-scientific world-view in a population is linked with support for democracy and law, let us make it clear that this does not mean that science alone – in a technical sense – is what ought to be taught in the schools. In fact, the inculcation of the rational-scientific world-view has always been accomplished through the broader curriculum, generally referred to as the 'liberal arts and sciences' course of study.

That is, rational-minded thinking is best engendered by teaching rational philosophy, historical analysis, literary and art criticism, pure mathematics and the pure sciences. This must be made clear, because the teaching of technical and applied sciences, without the liberal arts and sciences curriculum, could lead toward too narrow a world-view to sustain democracy. Let us look more closely at this.

TECHNOLOGICAL SPECIALIZATION AS A PROBLEM FOR DEMOCRACY

There are those intellectuals, such as Marcuse[8] and Arendt,[9] who worry that modern education will not engender an 'educated citizenry'. Rather, they fear that modern education will become so specialized, so linked to technological achievements, that it could lead to 'one-dimensional' men and women, who, as 'soulless experts', yield to 'heartless bureaucrats', who abdicate their citizen's responsibilities.

These fears are real. If the schools become technical training institutes, and lose their philosophical, scientific, and humanistic grounding, the democratic world-view could disappear, replaced by a new view of nature and society wherein hierarchies of 'technocrats' produce and distribute a high-technology abundance, while modern humans trade their freedom for the high-tech 'toys' disgorged by the modern economy.

It is heartening, therefore, that universities the world over are requiring that students take philosophy, history, literature, music, science, mathematics and languages – no matter what their technologically-oriented career might be later on. The hope is that through the continuation of the liberal arts and pure sciences curriculum, the high-technology economy will come to undergird a high-technology democracy, rather than a technological bureaucracy as in *Brave New World*, or worse, a technological totalitarianism, like that in *1984*.

Further, although the rational-scientific world-view itself, inadvertently engenders the mind-set for law and democracy – even where it is taught within an authoritarian context, such as in China or the former Soviet Union – it is still necessary, as Dewey alerted us[10] to teach about democracy and law directly, in order to ensure that a given population comes to conceive of itself as a citizenry with definite constitutionally guaranteed rights, and humanistic responsibilities.

EDUCATION ABOUT DEMOCRACY AND LAW: 'CIVICS'

First, we made it clear that a broad, general education was necessary for a majority of the citizenry if democracy as a political system was to flourish. Secondly, we made it clear that the inculcation of the rational-scientific world-view was critical for democracy in that it provides the basis for a citizen's acceptance of rational-legal guidelines for political action. Thirdly, we emphasized that the rational-scientific world-view is taught through a whole range of rational-analytic subjects and not through specialized, technological scientific studies alone.

Now, it must be emphasized that the content of democracy itself must be studied as part of the general educational programme in democratic societies. That is, what we used to call 'civics', or citizen's education, should be taught from the earliest grade-school years right through to university level.

Both the content and the process of democracy must be taught. That is, one teaches young people what democracy is all about: literally, the rules of the political process, and the structure of democratic government. In George Herbert Mead's terms, we socialize the young to democratic self-perception, by teaching them the rules of the game, and then encouraging them to play the game.[11]

To be explicit: within the civics courses, democratic political systems are explained and analysed. The European parliamentary system, for instance, would be compared to the American presidential system. The British two-party system could be compared to the European multi-party process. The direct democracy of ancient Greece or the Swiss cantons can be compared with modern representative democracy in the larger nation-states. English common law, with its unique evolution, can be compared with the Roman Law traditions of France and Germany.

Extending beyond European traditions, the democratic assemblies and clan councils of pre-literate tribes can be studied.[12] Biblical sources, for instance, in the Book of Judges,[13] are similar to those now being deciphered from Sumerian cuneiform tablets,[14] while the pre-literate democracy of the Iroquois Confederation[15] was so impressive that American colonists compared it to their 'articles of confederation' (even borrowed from it).

Precious democratic concepts should also be taught in the civics courses. Ideals such as free speech and free press, the limitation of the tenure and power of political leaders, and the right to electoral participation and toleration of minority viewpoints – these ideals are not as 'self-evident' as the American founders thought – they have to be carefully taught.[16]

In general, then, a curriculum can be developed which teaches the forms and functions of legal democracy, as well as its historical roots.

Along with the content of democracy, students can also be taught the processes. That is, they can be encouraged to participate in democratic programmes. Mock elections and moot courts become the vehicles through which young people are further socialized to the rules of the legal-democratic political game.

Mock elections can take many different forms: during actual national or local elections, for instance, the students could be organized to cast their votes as if they were eligible citizens. The tally of the results then can be made available to the students so that they could see whether their vote agreed with the actual vote. Students can also hold mock political conventions, or mock primaries, wherein they pick candidates for the elections. And, of course, students can hold actual elections for school offices.

As for courts, moot court, wherein students learn to bring cases, try case, judge and sit as jurors, has proven quite successful in explaining law and the court systems to young people. Along with this, some schools have held active courts, where student offenders were brought before elected student jurors who set penalties for minor offences in the schools.

The process of legal democracy, then, becomes as important as the content in inculcating a citizen's active, rational attitude toward political responsibilities and legal justice.

Finally, in this world of the mass media, television may become a helpful device in establishing the democratic ethos. Through the interactive television systems of the near future, 'television town-meetings' as we have suggested, may become possible, wherein citizens will become able to talk to their representatives, or participate in 'instant referenda' through computer-television hookups.

Since today's children – all over the world – are pre-socialized to this electronic world through their video-game systems (at home or in the public arcades) and through home or school computers, 'television town-meetings', held in the schools, may become a wonderful way to familiarize children with democratic processes and content – and encourage them to be active, rather than passive, participants in the new world of television politics.

Thus, instead of technological totalitarianism, like that of *1984*, the future telecommunications systems could produce technological democracy.

TOWARDS A LEGAL-DEMOCRATIC WORLD

Much of the world's population is still impoverished, illiterate, and linked to magico-religious world-views supporting traditional authority structures. However, the world is changing fast. How many theorists predicted the wave of democracy and consumer-demand that crashed across Eastern Europe and Asia and swept away communism?[17] Today's world culture is being carried to the younger generations by the mass media, and, it not only engenders consumerism, rock music, and high technology, but also demands for free expression, political participation, and legal protections.

If we take these new desires seriously, and adopt a Jeffersonian attitude, then the establishment of an educated citizenry with stable modern career choices becomes the basis from which a world democratic citizenship could emerge.

The world goals become: (1) a universal education system; (2) an international university system; (3) a liberal arts and sciences curriculum undergirding a rational-legitimacy system; (4) the civics curriculum, teaching democracy and law specifically; (5) the expansion of the modern high-technology industrial capitalist economy as a world system; (6) – lest we forget – the institutionalization of a legal-representative political system.

Establishing a universal educational system is easier than establishing a modern economy or stabilizing the legal-democratic processes during transitional phases of development. Of course, developing nations will be plagued by instability, but they should proceed with the educational processes and economic programmes. A nation such as India, for instance, though it still has a majority poor, has such a large, well-educated, rational-minded middle class, that at least a semblance of legal-democratic stability has emerged, and the educated middle class is, itself, helping to create the modern economic system which is undergirding the democratic process.

Finally, it should be emphasized that the steps outlined need not eradicate the national culture of the developing nation involved. This is not a 'westernization' process. Japan, for instance, has retained its culture indelibly, while instituting industrialization, education, science and democracy. Is France culturally like Britain or Germany? And didn't Spain, a so-called 'western' nation, resist democracy? And, now, without a cultural loss, Spain is becoming democratic.

Therefore, with Spain and Japan as models, Third World nations can become democratic, even where their traditional authority structure and culture seem loaded against it. Of course, the educational programme will take a long time to engulf a majority of the population, and even longer to engender a majority middle class. And the establishment of a viable, vibrant industrial economy is not easy in this world of developed superpower economies. The examples of Singapore and South Korea, as well as that of Brazil, show the potential for economic success among Third World nations. Yet these same nations still lag in terms of educational systems, nor has any of them so far created a stable legal-democratic political system.

Notes

INTRODUCTION

1. Ronald M. Glassman, *China in Transition: Communism, Capitalism, Democracy*, Westport, CT, Praeger, 1990.
2. Francis Fukayama, 'The End of History', *The Public Interest*, Summer, 1989.
3. Ronald M. Glassman, *The Middle Class and Democracy in Socio-Historical Perspective*, Leiden, The Netherlands, E.J. Brill, 1996.
4. Ibid.
5. Ibid.
6. Ibid.
7. Max Weber, *The Protestant Ethic and Spirit of Capitalism*, trans. Talcott Parsons, Beverly Hills, CA, Roxbury Press, 1995 (new edition, with an excellent introduction by Randall Collins).
8. Karl Polanyi, *The Great Transformation*, Boston, Beacon, 1976.
9. Benjamin Barber, *Strong Democracy*, New Brunswick, NJ, Rutgers University Press, 1982.
10. John Wicklein, *Electronic Nightmare: The New Communications and Freedom*, New York, Viking, 1981.
11. Robert Reich, *The Transnational World Economy*, Boston, Beacon, 1990.
12. Jean Jacques Rousseau, *The Social Contract*, Harmondsworth, Penguin Classics, 1983.
13. Robert Dahl, *After the Revolution*, New Haven, Yale University Press, 1972.
14. Benjamin Barber, *Strong Democracy*.
15. Ronald M. Glassman, William Swatos, Jr, Paul Rosen, *Bureaucracy Against Democracy and Socialism*, Westport, CT, Greenwood Press, 1986.
16. C. Wright Mills, *The Power Elite*, New York, Oxford University Press, 1962.
17. Wolfgang Mommsen, *The Age of Bureaucracy*, Chicago, University of Chicago Press, 1983.
18. Karl Marx, *Das Capital*, New York, International Publishers, 1936.
19. Adam Smith, *The Wealth of Nations*, London, Penguin, 1971.
20. Thorstein Veblen, *The Engineers and the Price System*, New York, Harcourt, Brace, and World, 1963.
21. Milton Friedman, *Capitalism and Democracy*, Chicago, University of Chicago Press, 1982.
22. John Maynard Keynes, *The General Theory*, New York, Harcourt, Brace, World, 1935.
23. Friedrich List, *The National System of Political Economy*, New York, Kelley, 1966.
24. Joseph Schumpeter, *Capitalism, Socialism, Democracy*, New York, Harper Torch Books, 1964.
25. John Kenneth Galbraith, *The New Industrial State*, Boston, Houghton Mifflin, 1966.

26. Daniel Bell, *The Coming of Post-Industrial Society*, New York, Basic Books, 1969.
27. Robert Reich, *The Trans-National World Economy*.
28. Gunnar Myrdahl, *The Challenge of Affluence*, New York, Harper and Row, 1959.
29. Robert Wade, *Governing the Market*, Princeton, NJ, Princeton University Press, 1990.
30. Ibid.
31. Ibid.
32. Ibid.
33. James Fallows, *Looking at the Sun*, New York, Pantheon, 1994.
34. Ibid.
35. Lester Thurow, *Head to Head*, Englewood Cliffs, NJ, Prentice Hall, 1990.
36. Reinhard Bendix, *Kings or People*, New York, Basic Books, 1981.
37. Karl Marx and Friedrich Engels, *The Communist Manifesto*, New York, International Publishers, 1936.
38. Aristotle, *The Politics*, New York, Oxford University Press, 1974, translated by Ernest Barker.
39. Ibid.
40. *The Oxford Study Bible*, New York, Oxford University Press, 1991 (see the Book of Judges and the Book of Kings).
41. Henri Frankfort, *Kingship and the Gods*, Chicago, University of Chicago Press, 1955.
42. Dun J. Li, *The Ageless Chinese*, New York, Charles Scribner's, 1965.
43. Ronald M. Glassman, *Democracy and Despotism in Primitive Societies*, New York, Associated Faculty Press, 1980.
44. Max Weber, *Economy and Society*, translated by Gunther Roth and Claus Wittich, New York, Academic Press, 1976.
45. Noam Chomsky, *The New Mandarins*, New York, Pantheon, 1975.
46. C. Wright Mills, *The Power Elite*, New York, Oxford University Press, 1962.
47. Wolfgang Mommsen, *The Age of Bureaucracy*, New York, Harper and Row, 1974.
48. *The Oxford Study Bible* (Exodus).
49. Friedrich Hayek, *The Road to Serfdom*, Chicago, University of Chicago Press, 1948.
50. Karl Popper, *The Open Society and Its Enemies*, Princeton, NJ, Princeton University Press, 1944.
51. Kevin Phillips, *The Politics of Rich and Poor*, New York, Random House, 1989.
52. Lester Thurow, *Head to Head*, Englewood Cliffs, NJ, Prentice Hall, 1990.
53. Robert Reich, *The Trans-National World Economy*.
54. Lester Thurow, *Head to Head*.
55. James Fallows, *Looking at the Sun*.
56. *'Ethnic Conflict in the Soviet Union:' A Symposium*, Baltimore, Johns Hopkins University Press, in *The Journal of Democracy*, Fall, 1996, No. 18.
57. James Fallows, *Looking at the Sun*.

Part I

1 HIGH-TECHNOLOGY INDUSTRIAL CAPITALISM AS A NEW MODE OF PRODUCTION

1. Berle and Means, *The Modern Corporation and Private Property*.
2. James Burnham, *The Managerial Revolution*, New York, Day, 1941.
3. Galbraith, *The New Industrial State*.
4. Bell, *The Coming of Post Industrial Society*.
5. Myrdahl, *Challenge to Affluence*.
6. Joseph Schumpeter, *Capitalism, Socialism, Democracy*, New York, Harper Torch Books, 1964.
7. Keynes, *The General Theory*.
8. John Kenneth Galbraith, *Economics in Perspective*, Boston, Houghton Mifflin, 1988. (Galbraith mentions the Swedish economists.)
9. Bell, *Post Industrial Society*.
10. Glassman, *China in Transition*.
11. Reich, *Trans-National World Economy*.
12. Veblen, *The Engineers and the Price System*.
13. Thorstein Veblen, *Imperial Germany*, New York, Viking Press, 1963.
14. Friedrich List, *The National System of Political Economy*, New York, Kelley, 1966.
15. Karl Marx and Friedrich Engels, *The Paris Commune*, ed. Hal Draper, New York International Books, 1952.
16. Max Weber, *Economy and Society*, ed. Claus Wittich and Gunther Roth, New York, Bedminster Press, 1974.
17. Adam Smith, *The Wealth of Nations*, Harmondsworth, Penguin Classics, 1982.
18. Lester Thurow, *Head to Head*, Englewood Cliffs, Prentice-Hall, 1990.
19. Robert Wade, *Governing the Market*, Princeton, NJ, Princeton University Press, 1990; Chalmers Johnson, *MITI*, Stanford, CA, Stanford University Press, 1982 and James Fallows, *Looking at the Sun*, New York, Pantheon Books, 1994.
20. Thurow, *Head to Head*.
21. Ibid.
22. Friedman, *Capitalism and Democracy*.
23. Galbraith, *New Industrial State*.
24. Thornstein Veblen, *The Theory of Business Enterprise*, Boston, Houghton Mifflin, 1953.
25. Galbraith, *New Industrial State*.
26. Kevin Phillips, *Arrogant Capital*, New York, Random House, 1995.
27. Ronald M. Glassman, *The Middle Class and Democracy in Socio-Historical Perspective*, Leiden, The Netherlands, E.J. Brill, 1996. See Glassman, Swatos and Kivisto, *For Democracy*, Westport, CT, Greenwood Press, 1953.
28. Burnham, *The Managerial Revolution*.
29. Noam Chomsky, *The New Mandarins*, New York, Pantheon Books, 1975.
30. Burnham, *The Managerial Revolution*.
31. Berle and Means, *The Modern Corporation and Private Property*.
32. Milovan Djilas, *The New Class*, Oxford, Oxford University Press, 1956.
33. Thurow, *Head to Head*.

34. Ibid.
35. Joel Kurtzman, *The Death of Money*, Boston, Little Brown, 1995, p. 11.
36. Ibid., p. 50.
37. Ibid., p. 52.
38. Ibid., p. 53.
39. Ibid., p. 16.
40. Ibid., p. 17.
41. Ibid., p. 18.
42. Ibid., p. 19.
43. Berle and Means, *The Modern Corporation and Private Property*.
44. Joel Kurtzman, *The Death of Money*, p. 137.

2 THREE MODELS OF HIGH-TECHNOLOGY INDUSTRIAL CAPITALISM

1. Max Weber, *The Protestant Ethic and Spirit of Capitalism*, Beverley Hills, Roxbury Press, 1996; see the excellent introduction by Randall Collins.
2. R.H. Tawney, *Religion and the Rise of Capitalism*, New York, Mentor Books, 1947.
3. Adam Smith, *The Wealth of Nations*, New York, Penguin Classics, 1982.
4. Weber, *Protestant Ethic*; Tawney, *Religion and the Rise of Capitalism*.
6. Charles Dickens, *Oliver Twist*, London, D.C. Heath, 1947.
7. John Calvin and William Tyndale, in Hans J. Hillerbrand, New York, Harper and Rowe, 1968.
8. Hayek, *The Road to Serfdom*.
9. Keynes, *The General Theory*.
10. Robert Heilbroner, *The Crisis of Vision in Modern Economic Thought*, Cambridge, Cambridge University Press, 1995.
11. Keynes, in Glassman, *Democracy and Equality*, New York, Praeger, 1983.
12. Aristotle, *Politics*, Barker translation, Oxford, Oxford University Press, 1958.
13. Max Weber, *The City*, Glencoe, IL, Free Press, 1962.
14. List, *National Political Economy*.
15. Veblen, *Imperial Germany*; Thurow, *Head to Head*.
16. Vatro Murvar and Ronald M. Glassman, *Max Weber's Political Sociology*, Westport, CT, Greenwood Press, 1980; see Vatro Murvar, 'On Weber's Prediction for Russia in 1918'.
17. Thurow, *Head to Head*.
18. James Fallows, *Looking at the Sun*.
19. Schumpeter, *Capitalism, Socialism, Democracy*.
20. Veblen, *Engineers*.
21. Thurow, *Head to Head*.
22. Ibid.
23. Seymour Martin Lipset, *America's Exceptionalism*, New York, W.W. Norton, 1996.
24. Donald Rowat, *The Ombudsman*, Lanham, MD, University Press of America, 1985.
25. Thurow, *Head to Head*.
26. Fallows, *Looking at the Sun*; Robert Wade, *Governing the Market*; Chalmers Johnson, *MITI*.

27. Max Weber, *The Religion of China*, Glencoe, IL, Free Press, 1962.
28. Weber, *Protestant Ethic*.
29. Fallows, *Looking at the Sun*; Wade, *Governing the Market*.
30. Metin Heper, *Bureaucracy of the State*, Westport, CT, Greenwood Press, 1985.
31. Ibid.
32. R.H. Tawney, *The Agrarian Problem in the Sixteenth Century*, New York, Harper Torch Books, 1967.
33. Fallows, *Looking at the Sun*; Wade, *Governing the Market*.
34. Fallows, *Looking at the Sun*; Wade, *Governing the Market*.
35. Fallows, *Looking at the Sun*.
36. Ibid.
37. Ibid.
38. Ibid.
39. Ibid.
40. Ibid.
41. Ibid.
42. Ibid.
43. S.N. Eisenstadt, *From Generation to Generation*, New York, Random House, 1971.
44. Glassman, Swatos, Rosen, *Bureaucracy Against Democracy and Socialism*.
45. E. Digby Baltzel, *The Protestant Establishment*, New York, Free Press, 1972.
46. G. William Domhoff, *Who Rules America*, Garden City, NY, Harper Torch Books, 1965.
47. C. Wright Mills, *The Power Elite*, New York, Oxford University Press, 1962.
48. Robert Dahl, *Who Governs*, New Haven, Yale University Press, 1963.
49. Robert Dahl, *After the Revolution*, New Haven, Yale University Press, 1972.
50. John Locke, *On Civil Government*, Harmondsworth, Penguin Classics, 1988; James Madison, Alexander Hamilton, *The Federalist Papers*, New York, Mentor Books, 1946.
51. Hal Draper (ed.), *The Paris Commune* (Marx and Engels), New York International Publishers, 1947.
52. Nicos Poulantzas, *Political Power and Social Classes*, New York, Schocken, 1973.
53. Draper (ed.), *Paris Commune*.
54. Wade, *Governing the Market*.
55. Ibid.
56. Ibid.
57. Ibid.
58. Ibid.

3 THE NEW CLASS STRUCTURE ENGENDERED BY THE HIGH-TECHNOLOGY ECONOMY, THE BUREAUCRATIC STATE AND THE SERVICE SECTOR

1. Lipset, *American Exceptionalism*.
2. Thurow, *Head to Head*.
3. Phillips, *Arrogant Capital*.
4. Christopher Lasch, *The Revolt of the Elites*, New York, Norton, 1996.

5. John Kenneth Galbraith, *A Journey Through Economic Time*, Boston, Houghton Mifflin, 1995.
6. Glassman, Swatos, Rosen, *Bureaucracy Against Democracy and Socialism*.
7. Ibid.
8. Burham, *Managerial Revolution*.
9. Berle and Means, *Modern Corporation and Private Property*.
10. Djilas, *The New Class*.
11. Glassman, *The Middle Class and Democracy in Socio-Historical Perspective*, Leiden, The Netherlands, E.J. Brill, 1996.
12. Arthur J. Vidich and Ronald M. Glassman, *Conflict and Control: The Challenge to Legitimacy of the Twentieth Century*, Beverley Hills, CA, Sage Publications, 1976.
13. Weber, *Religion of China*.
14. Mills, *Power Elite*.
15. Veblen, *Engineers and the Price System*.
16. Seymour Martin Lipset, *Political Man*, Baltimore MD, Johns Hopkins University Press, 1986.
17. Robert Dahl, *After the Revolution*, New Haven, CT, New York University Press, 1972.
18. C. Wright Mills, *White Collar*, New York, Oxford University Press, 1956.
19. Weber, *Protestant Ethic*.
20. Weber, *Economy and Society*.
21. Glassman, *China in Transition*.
22. Aldous Huxley, *Brave New World*, London, Heath, 1953.
23. Christopher Lasche, *The Culture of Narcissism*, New York, Warner Books, 1983.
24. George Orwell, *1984*, New York, Harcourt Brace, 1949.
25. Ronald M. Glassman, William Swatos Jr, Peter Kivisto, *For Democracy: The Noble Character and Tragic Flaws of the Middle Class*, Westport, CT, Greenwood Press, 1990.
26. Kevin Phillips, *The Politics of Rich and Poor*, New York, Random House, 1989.
27. Richard Cloward and A. Ohlin, *Delinquency and Opportunity*, New York, Columbia University Press, 1960.

Part II

4 THE NEW MIDDLE CLASS AS AN ARISTOTELIAN BASE FOR DEMOCRACY

1. Mills, *White Collar*.
2. Mills, *Power Elite*.
3. Glassman, *China in Transition*.
4. Henry Turner, *German Big Business and the Rise of Hitler*, New York, Oxford University Press, 1985.
5. Mills, *White Collar*.
6. Mills, *Power Elite*.
7. Glassman, Swatos, Kivisto, *For Democracy*.
8. Herbert Marcuse, *One Dimensional Man*, Boston, Beacon, 1964.

9. Hanna Arendt, *The Human Condition*, Chicago, University of Chicago Press, 1958.
10. Lasch, *The Culture of Narcissism.*
11. Marcuse, *One Dimensional Man.*
12. Theodore Roszack, *The Making of the Counter-Culture*, Los Angeles, University of California Press, 1995.
13. Lasch, *The Culture of Narcissism.*
14. Eric Fromm, *Escape from Freedom*, New York, Harper and Rowe, 1947.
15. Murvar and Glassman, *Max Weber's Political Sociology*; see Jose Casanova, 'Rationalization as the Unifying Concept in Weber's Work'.
16. Weber, *Economy and Society*, 'The Rational Forms of Action'.
17. Emile Durkheim, *Suicide*, New York, Free Press, 1968.
18. Arendt, *The Human Condition.*
19. Aristotle, *Ethics*, New York, Modern Library, 1954, translated by Jowett.
20. Plato, *The Republic*, New York, Penguin, 1986.
21. Thucydides, *The Pelopponesian War*, New York, Penguin, 1986. (See Pericles' Funeral Oration.)
22. Mills, *Power Elite.*
23. Ralph Nader and Mark Green, *Taming the Giant Corporation*. See also Ira Glasser, *Doing Good*, New York, Pantheon.
24. Donald Rowate, *Ombudsman.*
25. Robert Dahl, *After the Revolution.*
26. Marshall McLuhan, *The Medium is the Message*, New York, Random House, 1965 and *Understanding Media*, New York, American Library, 1964.

5 THE NEW MIDDLE CLASS AND LAW

1. Glassman, Swatos, Rosen, *Bureaucracy Against Democracy and Socialism.*
2. Weber, *Economy and Society*, 'Types of Legitimate Authority (Domination)'.
3. David Hume, *Moral and Political Philosophy*, New York, Columbia University Press, 1963.
4. Jean Jacques Rousseau, *The Social Contract*, New York, Modern Library, 1951.
5. Glassman, Swatos, Rosen, *Bureaucracy Against Democracy and Socialism.*
6. Ronald M. Glassman and William Swatos Jr, *Charisma, History, and Social Structure*, Westport, CT, Greenwood Press, 1981.
7. Wolfgang Mommsen, *The Age of Bureaucracy*, New York, Harper and Rowe, 1974.
8. Robert Dahl, *After the Revolution.*
9. Ira Glasser, *Doing Good*, New York, Pantheon Books, 1977.
10. Vidich and Glassman, *Conflict and Control.*
11. Mommsen, *The Age of Bureaucracy.*
12. Vidich and Glassman, *Conflict and Control.*
13. Glasser, *Doing Good*; Nader, Green, Seligman, *Taming the Giant Corporation.*
14. Jurgen Habermas, *Legitimation Crisis*, Boston, Beacon, 1976.
15. Arisotle, *Politics.*
16. Ibid.
17. Ibid.

18. Thomas Hobbes, *Leviathan*, Harmondsworth, Penguin Classics, 1984.
19. Eduard Bernstein, *Cromwell and Communism*, New York, Kelley, 1963; Philip Taylor, *The Origins of the English Civil War*, D.C. Heath, 1961; Christopher Hill, *God's Englishman*, New York, Dial Press, 1970.
20. Hobbes, *Leviathan*.
21. Aristotle, *Politics*; Polybius, Bk VI, *The Histories*; Von Fritz, *The Theory of the Mixed Polity*, New York, Columbia University Press, 1954.
22. Hobbes was quoted as saying 'Aristotle was the worst teacher that ever was.' *Leviathan, Behemoth*.
23. Hobbes, *Leviathan*.
24. Ibid.
25. Ibid.
26. Von Fritz, *Theory of the Mixed Polity*.
27. Hobbes, *Leviathan*.
28. Ibid.
29. Locke, *On Civil Government*.
30. Ibid.
31. Ibid.
32. Ibid.
33. Niccoló Machiavelli, *The Discourses*, New York, Everyman Edition, 1954.

6 MAINTAINING THE MIDDLE-CLASS MAJORITY ON THE HIGH-TECHNOLOGY INDUSTRIAL CAPITALIST BASE

1. Aristotle, *Politics*.
2. Ronald M. Glassman, *The Political History of Latin America*, New York, Funk and Wagnalls, 1969.
3. Glassman, *The Middle Class and Democracy in Socio-Historical Perspective* (chapter on 'communism').
4. Aristotle, *Politics*.
5. Ibid.
6. David Bazelon, *The Paper Economy*, New York, Random House, 1962.
7. Louie Kelso and Mortimer Adler, *The Capitalist Manifesto*, New York, Random House, 1955.
8. Phillips, *Arrogant Capital*.
9. Phillips, *Politics of Rich and Poor*.
10. Keynes, *General Theory*.
11. Thurow, *Head to Head*.
12. Keynes, *General Theory*; see also, Glassman, *Democracy and Equality*.
13. Aristotle, *Politics*.
14. Ibid.
15. Ibid.
16. Ibid.
17. Ibid.
18. Ibid.
19. Thurow, *Head to Head*; Phillips, *Arrogant Capital*.
20. Friedrich Hayek, *The Constitution of Liberty*, Chicago, University of Chicago Press, 1983.

Part III

7 BUREAUCRACY AS A DESPOTIC SYSTEM OF DOMINATION

1. Glassman and Swatos, *Charisma, History, and Social Structure.*
2. Ibid.
3. Eli Chinoy, *Sociology*, New York, Random House, 1961 (chapter on 'Bureaucracy).
4. Weber, *Religion of China.*
5. Mills, *Power Elite.*
6. Henri Frankfort, *Kingship and the Gods*, Chicago, University of Chicago Press, 1955.
7. Weber, *Religion of China.*
8. Egyptian *Book of the Dead*, and *Wisdom Literature of Babylon*, any editions.
9. Vidich and Glassman, *Conflict and Control.*
10. Hanna Arendt, *The Origins of Totalitarianism.*
11. Nader and Green, *Taming the Giant Corporation.*
12. The American Civil Liberties Union, 43rd Street, New York City, has pamphlets available on corporate dossier files.
13. J. Edgar Hoover had infamous dealing with J.F.K. and Martin Luther King, Jr.
14. Leonid Brezhnev used less violence than Stalin, but still kept firm order.
15. David Rothman, *Willowbrook*, New York, Schocken Books, 1973.
16. Andrew Stein, the city council president of New York City, uncovered the nursing home scandals of the 1960s.
17. During the ousting of Salvadore Allende in Chile (1960s), US corporate executives became involved with the counter-revolutionary violence.
18. Hanna Arendt, *Eichmann in Jerusalem*, New York, Viking, 1964.
19. Most of the CIA directors were members of the 'protestant establishment', Allen Dulles being the best-known, along with George Bush, who later became President of the USA.
20. Glassman, Swatos, *Charisma, History, Social Structure.*
21. Mills, *Power Elite.*
22. Robert K. Merton, 'Bureaucratic Personality', in *Social Theory and Social Structure*, New York, Columbia University Press, 1951.
23. Ralph Nader, *Unsafe at Any Speed*, New York, Grossman, 1972.
24. Arendt, *Eichmann in Jerusalem.*
25. Norman Podhoritz and Irving Kristol are well-known neo-conservatives of the 1970s. *Commentary Magazine* became a neo-conservative forum.
26. Noam Chomsky, *The New Mandarins.*
27. Weber, *Religion of China.*

8 DOES THE EMPIRE ANALOGY HOLD, OR ARE CRITICAL DIFFERENCES EMERGING?

1. Habermas, *Legitimation Crisis.*
2. Lewis Mumford, *The Myth of the Machine*, New York, Harcourt Brace & World, 1967.
3. Plato, *Republic*; Thomas More, *Utopia*, Harmondsworth, Penguin Classics, 1982.

4. Aristotle, *Politics*, section on 'proportional equality'.
5. Hayek, *The Constitution of Liberty*.
6. Glassman, Swatos, Rosen, *Bureaucracy Against Democracy and Socialism*.
7. Robert Dahl, *After the Revolution*; Nader & Green, *Taming the Giant Corporation*.
8. Rowat, *Ombudsman*.
9. Mills, *White Collar*.
10. Ronald M. Glassman and Robert J. Antonio, *A Weber–Marx Dialogue*, Lawrence, Kansas, University of Kansas Press, 1981. See the paper by Gunther Roth, 'Weber and Marx's Predictions 100 years after the death of Marx'.
11. Robert Michels, *Political Parties*, Glencoe, IL, Free Press, 1958.
12. Max Weber, *The Sociology of Religion*, Glencoe, Ill., Free Press, 1964.
13. Karl Marx, *Communist Manifesto*, New York, International Books, 1947.
14. Lasch, *Culture and Narcissism*.
15. Weber, *Economy and Society*, section on 'charismatic authority'.
16. Vance Packard, *The Hidden Persuaders*.
17. Marshall MacLuhan, *Understanding Media*.
18. Huxley, *Brave New World*.
19. Orwell, *1984*.

9 POLITICAL CULTURE AGAINST DEMOCRACY

1. Lucien Pye, *The Dynamics of Chinese Politics*, Cambridge, MA, Harvard University Press, 1981.
2. Samuel Huntington, *The Third Wave*, Cambridge, MA, Harvard University Press, 1988.
3. Ibid.
4. Seymour Martin Lipset, *Political Man*, Baltimore, MD, Johns Hopkins University Press, 1980.
5. Plato, *Republic*.
6. Ibid.
7. Aristotle, *Politics*.
8. Ibid.
9. Niccoló Machiavelli, *The Discourses*, Harmondsworth, Penguin Classics, 1982.
10. Thomas Paine, *The Rights of Man*, New York, Everyman Edition, 1953; see also *The Declaration of Independence* of the USA.
11. Paine, *The Rights of Man*.
12. Locke, *On Civil Government*; John Stuart Mill, *On Representative Government*, Harmondsworth, Penguin Classics, 1982.
13. Jean Jacques Rousseau, *Emile*, New York, Mentor Books, 1948.
14. Thucydides, *Pelopponesian War*, New York, Modern Library, 1955.
15. Edmund Burke, *Reflections on the Revolution in France*, London, Robinson, 1791.
16. Johann Gottlieb Fichte, *Selected Works*, Berlin, DeGruyter, 1965.
17. Miguel de Unamuno, *Selected Works*, Princeton, NJ, Princeton University Press, 1967.
18. *Max Weber's Political Sociology*, eds Murvar and Glassman, Westport, CT, Greenwood Press, 1980.

19. Ronald M. Glassman, *Political History of Latin America*, New York, Funk and Wagnalls, 1969.
20. Ibid.
21. Graeb Memorandum, in the *Columbia Source Books*, New York, Columbia University, 1948.
22. Ernest Hemingway, *For Whom the Bell Tolls*, New York, Harper and Rowe, 1952.
23. Glassman, *The Middle Class and Democracy in Socio-Historical Perspective; For Democracy: The Noble Character and Tragic Flaws of the Middle Class*.
24. Robert Wade, *Governing the Market*.
25. James Fallows, *Looking at the Sun*.
26. Glassman, *China in Transition*.
27. Glassman and Swatos, *Charisma, History, and Social Structure*.
28. Weber, *Protestant Ethic*; see Randall Collins' introduction in the Roxbury Press edition, 1996.
29. Lipset, *Political Man*.
30. Weber, *Economy and Society*; section on 'cesaro-papism'.
31. Machiavelli, *The Discourses*.
32. Lipset, *Political Man*.
33. 'Liberation theology' was popular in Latin America during the 1970s and 1980s.
34. Robin Wright, 'Two Visions of Reformation: Soroush and al-Ghannouchi', in *The Journal of Democracy*, April 1996, Vol. 7, No. 2, Baltimore, MD, Johns Hopkins University Press.
35. *The Koran*, Penguin Classics, Harmondsworth, 1982.
36. Francis Peters, *Aristotle and the Arabs*, New York, New York University Press, 1968; Francis Peters, *The Harvest of Hellenism*, New York, Simon and Schuster, 1971.
37. Peters, *Aristotle and the Arabs*.
38. Bernard Lewis, 'A Historical Overview on Islam and Liberal Democracy', in *Journal of Democracy*, April 1996, Vol. 7, No. 2.
39. *The Koran*.
40. Robin Wright, 'Two Visions of Reformation: Soroush and al-Ghannouchi', *Journal of Democracy*, April 1996, Vol. 7, No. 2.
41. Francis Peters, *Aristotle and the Arabs*; *The Harvest of Hellenism*.
42. Robin Wright, 'Two Visions'.
43. Confucius, *The Analects*, Harmondsworth, Penguin Classics, 1982.
44. Plato, *Republic*.
45. Confucius, *The Analects*.
46. Weber, *Religions of China*.
47. Ibid.
48. Glassman, *China in Transition*.
49. Weber, *Religions of China*.
50. Ibid.
51. J.S. Mill, *On Representative Government*.
52. Lee Kwan Kue, founder of modern Singapore.
53. The 'old men' of the Chinese communist party insisted that to give in to the students' demands for free speech and elections would create 'disorder' in China.

54. James Fallows, *Looking at the Sun.*
55. Herodotus, *The Histories*, New York, Modern Library Edition, 1953.
56. Tacitus, *The Annals of Imperial Rome*, Harmondsworth, Penguin Classics, 1982.
57. Bible, *Oxford Study Bible*, New York, Oxford University Press, 1986.
58. Ibid.; see also, Carol A. Newsom and Sharon H. Ringe (eds), *The Women's Bible Commentary*, Louisville, KY, Westminster, John Knox Press, 1987.
59. Weber, *Religions of China.*
60. Vidich and Glassman, *Conflict and Control.*
61. Fallows, *Looking at the Sun.*
62. Wade, *Governing the Market.*
63. Francis Fukayama, 'Confucius and Democracy', in *Journal of Democracy*, April 1995, Vol. 6, No. 2.
64. Ibid.
65. Ibid.
66. Andrew Nathan, *Democracy on Taiwan*, New York, Columbia University Press, 1992.
67. Alice H. Amsden, *Asia's Next Giant*, New York, Oxford University Press, 1989.
68. Glassman, *Political History of Latin America*; Juan Linz, 'Latin American Democracy', *Journal of Democracy*, October, 1995, Vol. 4, No. 2.
69. Glassman, *The Middle Class and Democracy in Socio-Historical Perspective.*
70. Glassman and Swatos, *For Democracy: The Noble Character and Tragic Flaws of the Middle Class.*

Part IV

10 THE EXTENSION OF LEGAL AUTHORITY OVER PUBLIC AND PRIVATE BUREAUCRACIES

1. Noam Chomsky, *New Mandarins.*
2. Robert Dahl, *After the Revolution.*
3. Ralph Nader and Mark Green, *Taming the Giant Corporation.*
4. Milton Friedman, *Capitalism and Democracy.*
5. Ira Glasser, *Doing Good*, New York, Pantheon, 1982.
6. Ibid.
7. Nader and Green, *Taming the Giant Corporation.*
8. Ibid.
9. Ibid.
10. Ibid.
11. Ibid.
12. Ibid.
13. Ibid.
14. Ibid.
15. Glasser, *Doing Good.*
16. Ibid.
17. Ibid.
18. Ibid.

19.   Ibid.
20.   Ibid. See also Erving Goffman, *Asylums*, Glencoe, IL, Free Press, 1962.
21.   Glasser, *Doing Good*.
22.   Rowat, *Ombudsman*.
23.   Ibid.
24.   Ibid.; see also Papers of the American Civil Liberties Union, 'Corporate Dossiers and the Invasion of Privacy', 1970. Ira Glasser, ACLU, 43rd Street, New York City.
25.   Rowat, ibid.
26.   Ibid.

11   MASS-MEDIATED DIRECT DEMOCRACY

1.   Robert Michels, *Political Parties*.
2.   Gunther Roth, 'Weber and Marx's Predictions One Hundred Years after Marx's Death', in *A Weber–Marx Dialogue*, eds Ronald M. Glassman and Robert J. Antonio, Lawrence, Kansas, Kansas University Press, 1985.
3.   Robert Dahl, *Who Governs?*
4.   David Riesman, *Lonely Crowd*, Cambridge, MA, Harvard University Press, 1951.
5.   Campaign-financing legislation is being studied in the USA, Britain, and the EU – especially as this relates to television-time purchasing.
6.   Ibid.
7.   John Wicklein, *Electronic Nightmare: The New Communications and Freedom*, New York, Viking, 1981.
8.   Ibid.
9.   Ibid.
10. Ibid.
11. Ibid.
12. Ibid.
13. Ibid.
14. Ibid.
15. Ibid.
16. Ibid.
17. Ronald M. Glassman, *Democracy and Despotism in Primitive Societies*, New York, Associated Faculty Press, 1978.
18. Polybius, *The Histories*, Bk VI, New York, Modern Library Edition, 1957.
19. Benjamin Barber, *Strong Democracy*, New Brunswick, Rutgers University Press, 1982.
20. Ibid.
21. Ibid.
22. Ibid.
23. Ibid.
24. Polybius, *The Histories*, Bk VI.
25. James Madison and Alexander Hamilton, *The Federalist Papers*, New York, Harper Torch Books, 1958.
26. Barber, *Strong Democracy*.
27. Aldous Huxley, *Brave New World*.

28. George Orwell, *1984*.
29. ACLU papers, 'Corporate Dossiers and the Invasion of Privacy', Ira Glasser, ACLU, 43rd Street, New York City.
30. Hobbes, *Leviathan*.
31. Weber, *Economy and Society*, section on 'the state'.
32. Locke, *On Civil Government*.
33. John Wicklein, *Electronic Nightmare*.
34. Ibid.
35. Ibid.
36. Ibid.
37. Ibid.
38. Ibid.
39. Ibid.

12   EDUCATION FOR DEMOCRACY IN THE HIGH-TECHNOLOGY GLOBAL VILLAGE

1. Jose Ortega y Gasset, *The Revolt of the Masses*, New York, Macmillan, 1949.
2. Henri Frankfort, *Kingship and the Gods*.
3. Weber, *The Sociology of Religion*, section 'Science and Disenchantment'.
4. St Thomas Acquinas, *Summa Theologia*, Harmondsworth, Penguin Classics, 1982.
5. Emile Durkheim, *The Elementary Forms of Religious Life*, Glencoe, IL, Free Press, 1957.
6. Isaac Newton, 'On the Laws of Motion', in *Mathematical Principles*, Berkeley, University of California Press, 1960. See also Einstein's famous statement, 'I can't believe that God throws dice with men', commenting on Heisenberg's 'uncertainty principle'.
7. Weber, *Economy and Society*; section on 'legal authority'.
8. Hebert Marcuse, *One Dimensional Man*.
9. Hanna Arendt, *The Human Condition*.
10. John Dewey, *Democracy and Education*, New York, Macmillan, 1964.
11. George Herbert Mead, *Mind, Self, Society*, Chicago, University of Chicago Press, 1931.
12. Glassman, *Democracy and Despotism in Primitive Societies*; also Lewis Henry Morgan, *Ancient Society*, New York, Harcourt Brace, 1946.
13. *Bible*, Bk of Judges, Oxford Press edition.
14. Thorkild Jakobson, *Democracy in Sumeria*, Chicago, University of Chicago Press, 1953.
15. Lewis Henry Morgan, *The Iroquois League*, New York, World Publishing, 1941 and *Ancient Society*.
16. Preamble to the USA Constitution.
17. Samuel Huntington, *The Third Wave*, Norman, OK, University of Oklahoma Press, 1995.

Index